Biological Research on Industrial Yeasts

Volume I

Editors

Graham G. Stewart
Labatt Brewing Company
London, Ontario, Canada

Inge Russell
Labatt Brewing Company
London, Ontario, Canada

Ronald D. Klein
The Upjohn Company
Kalamazoo, Michigan

Ronald R. Hiebsch
The Upjohn Company
Kalamazoo, Michigan

CRC Press, Inc.
Boca Raton, Florida

Library of Congress Cataloging-in-Publication Data

Biological research on industrial yeasts.

"Symposium on the Biochemistry and Molecular Biology of Industrial Yeasts, Brook Lodge, Kalamazoo, Mich., October 25 to 30, 1985, sponsored jointly by the Upjohn Company, Kalamazoo and Labatt Brewing Company, Ltd., London, Ontario, Canada"--P.
Bibliography: p.
Includes index.
1. Yeast--Congresses. 2. Yeast industry--Congresses. I. Stewart, Graham G., 1942- II. Symposium on the Biochemistry and Molecular Biology of Industrial Yeasts (1985 : Kalamazoo, Mich.) III. Upjohn Company. IV. Labatt Brewing Company.
QR151.B565 1987 660′.62 87-838
ISBN 0-8493-4900-1 (set)
ISBN 0-8493-4901-X (v.1)
ISBN 0-8493-4902-8 (v.2)
ISBN 0-8493-4903-6 (v.3)

Direct all inquiries to CRC Press, Inc., 2000 Corporate Blvd., N.W., Boca Raton, Florida, 33431.

International Standard Book Number 0-8493-4900-1 (Set)
International Standard Book Number 0-8493-4901-X (Volume I)
International Standard Book Number 0-8493-4902-8 (Volume II)
International Standard Book Number 0-8493-4903-6 (Volume III)
Library of Congress Card Number 87-838
Printed in the United States

FOREWORD

One of the major goals of the representatives from Upjohn and Labatt Brewing Company, Ltd. in planning this symposium* was to bring together, in one forum, industrial and academic scientists. It was concluded that both the topic and timing of the symposium were appropriate for disproving several "myths" surrounding the relationship between academic and industrial research. These myths range from industrial research being solely applied, to the widely held feeling that the focus of academic and industrial research differ significantly. The organizers of the symposium wished to address these misconceptions in several ways. First, it appeared that the topics of interest were common both to academic and industrial researchers, and that both the title and content of the symposium should reflect the attempt to bridge the usually presumed "academic-industrial gap" in research interests. Next, the participants and their contributions should reflect the full range of research interests, from studies of the basic mechanisms of yeast molecular biology to application of yeast expression technologies in industries as diverse as brewing, oil production and pharmaceuticals. Finally, it was concluded that the state-of-the-art was of interest, and not review-type papers, and the participants were requested to prepare their presentations accordingly.

The results of the experiment, at least as measured by the anecdotal observations of both participants and organizers, indicated that a successful experiment has been carried out in essentially all respects. International participants from both industrial and academic laboratories engaged in lively presentations and debate throughout the symposium, and new, unpublished research results were at the core of most presentations. Furthermore, research results presented as posters received equal weight to oral presentations and are included as full manuscripts in this current publication.

It is a great pleasure, therefore, to thank both the sponsors of the symposium, The Upjohn Company and Labatt Brewing Company, Ltd., and the participants for a stimulating and scientifically important symposium. It is our hope that the proceedings of this symposium will not only reflect the exciting synergism that occurred during the symposium, but will also contribute substantially to the scientific literature in this area.

Ralph E. Christoffersen, Ph.D.
Vice President
Biotechnology and Basic Research Support
The Upjohn Company

* Symposium on the Biochemistry and Molecular Biology of Industrial Yeasts, Brook Lodge, Kalamazoo, Mich., October 25 to 30, 1985, sponsored jointly the The Upjohn Company, Kalamazoo and Labatt Brewing Company, Ltd., London, Ontario, Canada.

THE EDITORS

Graham G. Stewart, Ph.D., D.Sc., is Director of Research and Quality Control for the Labatt Brewing Company, Ltd., London, Ontario, Canada. He received his B.Sc. Hons. Microbiology and B.Sc. Hons. Biochemistry degrees from the University of Wales, his Ph.D. in Microbial Biochemistry from Bath University, and a D.Sc. from Bath University in 1983. Between 1967 and 1969, Dr. Stewart lectured in Biochemistry at the Portsmouth College of Technology. He joined the Labatt Brewing Company in 1969 and has worked, published, and lectured extensively on the biochemistry of yeast.

Dr. Stewart, as Chairman of the International Commission for Yeasts, organized the 5th International Symposium for Yeasts held in London, Ontario in July, 1980. He is presently Vice-Chairman of the International Commission for Yeasts and Secretary of the Biotechnology Commission of IUPAC. He is co-editor of the CRC Press journal, *Critical Reviews in Biotechnology,* is on the editorial board of the *Journal of Food and Microbiology* and of *Biotechnology Letters,* as well as being on the advisory board of *Microbiological Sciences.* In addition to editing a number of books, Dr. Stewart has published over 100 original papers, patents, and reviews.

Inge Russell, M.Sc., is Manager of the Microbiology/Genetics Section of the Research Department of the Labatt Brewing Company. She received her B.Sc. in Biology from the University of Western Ontario and her M.Sc. from the University of Strathclyde. She joined the Labatt Brewing Company in 1979 and has carried out research in the field of yeast genetics and biochemistry.

She is a co-editor of the CRC Press journal, *Critical Reviews in Biotechnology,* and is also on the editorial board of the *Journal of Industrial Microbiology.* In addition to editing a number of books, she has published over 40 original papers, patents, and reviews.

Ronald Klein, Ph.D., is a Research Scientist in Molecular Biology Research at The Upjohn Company, Kalamazoo, Michigan. Dr. Klein began his scientific career after completion of his B.A. in Humanities and M.A. in International Studies at Western Michigan University. He majored in Biology and Chemistry at Kalamazoo College, completed his Senior Research Project under Dr. Oldrich K. Sebek at The Upjohn Company, and was a Graduate Fellow in the Biochemistry Department at Wayne State University. Dr. Klein received his Ph.D. in Molecular Biology from the University of Wisconsin-Madison followed by postdoctoral study at the Massachusetts Institute of Technology with Dr. H. Gobind Khorana. He was a Research Scientist in the Biotechnology Division of the Phillips Petroleum Company before joining The Upjohn Company in 1984. Dr. Klein is carrying out research in the molecular biology and biochemistry of industrial yeasts.

Ronald R. Hiebsch, B.Sc., is a Research Associate in Molecular Biology Research at The Upjohn Company. Mr. Hiebsch completed his B.Sc. at Kansas State University where he worked on the mechanism of T-7 DNA replication. Mr. Hiebsch was a Research Assistant in the Department of Cell Biology and Anatomy at John Hopkins Medical School and later was a Research Associate in the Department of Biological Carcinogenesis at The Frederick Cancer Research Facility where he worked on the molecular biology of a C-typed retrovirus. His work at The Upjohn Company is focused on the biochemistry and molecular biology of industrial yeasts.

ACKNOWLEDGMENTS

The editors wish to acknowledge the support of The Upjohn Company and Labatt Brewing Company, Ltd., in the organization of the symposium and in the compilation of this monograph. Particular gratitude is given to Janice Riddell who painstakingly assembled the program and then compiled the original manuscripts of this document. Gratitude is also given to Patrick J. Delehanty, Kathy J. Hiestand, Laurie A. Tuinstra, Gurnam S. Gill, and Phillip G. Zaworski of The Upjohn Company, and to Chandra J. Panchal and Carl A. Bilinski of Labatt Brewing Company whose advise and assistance in the organization of the symposium was invaluable.

CONTRIBUTORS

Lars André
Biotechnical Laboratory
VTT
Espoo, Finland

A. Ben-Bassat
Cetus Corporation
Emeryville, California

L.M. Bennett
Research Laboratories
The Upjohn Company
Kalamazoo, Michigan

Charles Boone
Graduate Student
Department of Biology
McGill University
Montreal, Quebec, Canada

Richard Bozzato, Ph.D.
Division of Molecular Biology
Allelix Inc.
Mississauga, Ontario, Canada

Howard Bussey, Ph.D.
Professor
Department of Biology
McGill University
Montreal, Quebec, Canada

Bjorn E. Christensen
A/S De Danske Spritfabrikker
Copenhagen, Denmark

G.E. Cole
Cetus Corporation
Emeryville, California

R. Wayne Davies, Ph.D.
Division of Molecular Biology
Allelix Inc.
Mississauga, Ontario, Canada

Sergio Echeverrigaray
Escola Superior/Agricultura
Universidade de Sao Paulo
Piracicaba, Brazil

M. El-Kashab
Chemical and Biochemical Engineering
University of Western Ontario
London, Ontario, Canada

Seymour Fogel, Ph.D.
Professor
Department of Genetics
University of California, Berkeley
Berkeley, California

D.H. Gelfand
Cetus Corporation
Emeryville, California

Gurnam S. Gill, Ph.D.
Research Scientist
Department of Molecular Biology
The Upjohn Company
Kalamazoo, Michigan

Claes Gjermansen, M.Sc.
Department of Physiology
Carlsberg Laboratory
Copenhagen Valby, Denmark

Deirdre Greene, M.Sc.
Research Assistant
Department of Biology
McGill University
Montreal, Quebec, Canada

G.E. Guidoboni, BTech. CEng. MIChemE.
Senior Partner
GRC Consultants
Alton, Hants, England

Donald O. Hitzman, Ph.D.
Research Director
Geo-Microbial Technologies, Inc.
Ochelata, Oklahoma

M.J. Holland
Cetus Corporation
Emeryville, California

Steen Holmberg, Ph.D.
Department of Physiology
Carlsberg Laboratory
Copenhagen Valby, Denmark

James E. Hopper, Ph.D.
Department of Biological Chemistry
and the Cancer Research Center
The Milton S. Hershey Medical Center
Pennsylvania State University
Hershey, Pennsylvania

D. Inlow
Cetus Corporation
Emeryville, California

M.A. Innis
Cetus Corporation
Emeryville, California

R.M. Jones
Production Research Department
Labatt Brewing Company, Ltd.
London, Ontario, Canada

Morten C. Kielland-Brandt, Ph.D.
Professor
Department of Physiology
Carlsberg Laboratory
Copenhagen Valby, Denmark

Jonathan Knowles, Ph.D.
Research Professor
Biotechnical Laboratory
VTT
Espoo, Finland

Cletus P. Kurtzman, Ph.D.
Research Leader
Northern Regional Research Laboratory
USDA/ARS
Peoria, Illinois

Päivi Lehtovaara
Biotechnical Laboratory
VTT
Espoo, Finland

Susan J. Lolle, B.Sc.
Ph.D. Graduate Student
Department of Biology
McGill University
Montreal, Quebec, Canada

Argyrios Margaritis, Ph.D., P.Eng., F.C.I.C.
Professor
Department of Chemical and Biochemical Engineering
Faculty of Engineering Science
University of Western Ontario
London, Ontario, Canada

K.R. Marotti
Research Laboratories
The Upjohn Company
Kalamazoo, Michigan

P.C. McCabe
Cetus Corporation
Emeryville, California

Gary L. McKnight, Ph.D.
Senior Scientist
ZymoGenetics, Inc.
Seattle, Washington

J. McRae
Cetus Corporation
Emeryville, California

J.H. Meade
Cetus Corporation
Emeryville, California

Torsten Nilsson-Tillgren, Ph.D.
Institute of Genetics
University of Copenhagen
Copenhagen, Denmark

Stephen G. Oliver, Ph.D.
Director
Manchester Biotechnology Centre
UMIST
Manchester, England

Merja Penttilä, M.Sc.
Biotechnical Laboratory
VTT
Espoo, Finland

Jens G.L. Petersen, Ph.D.
Department of Physiology
Carlsberg Laboratory
Copenhagen Valby, Denmark

Inge Russell, M.Sc.
Production Research Department
Labatt Brewing Company, Ltd.
London, Ontario, Canada

Irma Salovuori
Biotechnical Laboratory
VTT
Espoo, Finland

L.K. Shay, Ph.D.
Senior Microbiologist
Research and Development
Provesta Corporation
Bartlesville, Oklahoma

Nigel A. Skipper, Ph.D.
Department of Molecular Biology
Allelix Inc.
Mississauga, Ontario, Canada

Graham G. Stewart, Ph.D.
Production Research Department
Labatt Brewing Company, Ltd.
London, Ontario, Canada

Rony Tal
Cetus Corporation
Emeryville, California

Flavio C.A. Tavares, Ph.D.
Professor
Department of Genetics
University de Sao Paulo
Piracicaba, Brazil

Derk te Bokkel, M.E.Sc.
Graduate Student
Chemical and Biochemical Engineering
University of Western Ontario
London, Ontario, Canada

Tuula Teeri
Biotechnical Laboratory
VTT
Espoo, Finland

David Y. Thomas, Ph.D.
Director
Molecular Genetics Division
Biotechnology Research Institute
National Research Council of Canada
Montreal, Quebec, Canada

Darrell R. Thomsen, B.S.
Research Associate
Department of Molecular Biology
The Upjohn Company
Kalamazoo, Michigan

Alan Upshall, Ph.D.
Senior Scientist
Zymogenetics, Inc.
Seattle, Washington

Thierry Vernet, Ph.D.
Research Officer
Biotechnology Research Institute
National Research Council of Canada
Montreal, Quebec, Canada

Daniel Vetter
Division of Molecular Biology
Allelix Inc.
Mississauga, Ontario, Canada

G.H. Wegner, Ph.D.
Vice President Research and Development
Provesta Corporation
Bartlesville, Oklahoma

Juliet W. Welch, M.A.
Staff Research Associate
Department of Genetics
University of California, Berkeley
Berkeley, California

V.P. Wittman
Cetus Corporation
Emeryville, California

Raymond Wong, Ph.D.
Division of Molecular Biology
Allelix Inc.
Mississauga, Ontario, Canada

Phillip G. Zaworski, M.S.
Biochemist
Molecular Biology Research
The Upjohn Company
Kalamazoo, Michigan

TABLE OF CONTENTS

Chapter 1

YEAST — THE PRIMARY INDUSTRIAL MICROORGANISM

Inge Russell, R. Jones, and Graham G. Stewart

TABLE OF CONTENTS

I. INTRODUCTION

Yeasts are, without a doubt, both quantitatively and economically the most important group of microorganisms commercially exploited by man. The total amount of yeast produced annually, including that formed during brewing, wine-making, and in distilling practices, is in the order of a million tonnes. This group of microorganisms is, by traditional agreement, limited to "fungi" in which the unicellular form is predominant and vegetative reproduction is usually, but not always, by budding (Figure 1). This group does not constitute a taxonomic entity, although it comprises a subdivision of narrowly related species. Compared to other major groups of microorganisms — algae, bacteria, and protozoa — the yeasts are represented by comparatively few genera and species. It is thought that there are only 500 species grouped into 60 genera, however, the currently accepted text on yeast taxonomy is 1082 pages long.[1]

Some biochemists and molecular biologists incorrectly employ the term yeast as synonymous with *Saccaromyces cerevisiae.* Although most of the biochemical and genetic studies have been conducted on this species, there is a rich variety of yeast genera and species that await biological investigation; indeed, the most exotic species may offer advantages for experimental study. Nevertheless, the genus *Saccharomyces* has often been referred to as "the oldest plant cultivated by man." Indeed, the history of beer, wine, and breadmaking, with the fortuitous use of yeast, is as old as the history of man himself; however, recent years have seen the transition of yeast from being solely the "workhorse" in the traditional food and beverage industry to being, in addition, one of the organisms of choice for gene manipulation involving the "new genetics".

It is the opinion of some biotechnologists that the future importance of this group of yeasts will decline due to the fact that they produce few secondary metabolites of commercial interest and that their ability to secrete quantities of protein is limited when compared to many mycelial fungi (this aspect is discussed in a subsequent paper in this volume.[2] However, many species of yeast (particularly of the genus *Saccharomyces*) are of GRAS (Generally Regarded As Safe) status and produce two very important primary metabolites — ethanol and carbon dioxide. The ethanol is used in both beverages and for industrial or fuel purposes. The carbon dioxide is employed for leavening in baked goods, for carbonation of beverages, as a solvent in its liquid state, and in the culturing of vegetables in greenhouses under controlled environmental conditions. In addition, there are a number of other important industrial uses for yeast, including cultures that have been genetically transformed to produce important nonyeast proteins and peptides, such as the antiviral protein interferon,[3] human serum albumin,[4] and the acid protease chymosin used in the milk-clotting step in cheese production.[5]

Yeasts have been, still are, and will be in the future, important model experimental organisms for the elucidation of the basic biochemical, metabolic, and genetic processes of living cells. *S. cerevisiae* and related species are currently favored as experimental organisms, because they are eukaryotes that can be cultured and analyzed genetically with similar ease as bacteria. The power of the yeast genetic system is a critical and important parameter. Cells grow stably as haploids, diploids, aneuploids, or polyploids. It is easier to obtain mutants in the haploid phase than in the diploid or polyploid cells of higher plants or animals. It is possible in the diploid phase of *S. cerevisiae* to determine whether mutants are dominant or recessive, to classify them into complementation groups, and to map their location in the genome. During the past 40 years, genetic studies have yielded a great deal of information on yeast genes — well over 350 loci have been mapped to 17 linkage groups.[6] In addition, considerable effort has been devoted to ways of manipulating and modifying the yeast genome employing classical techniques — mainly, mutation and hybridization. This knowledge has provided the essential background for gene cloning and host-vector development in *S.*

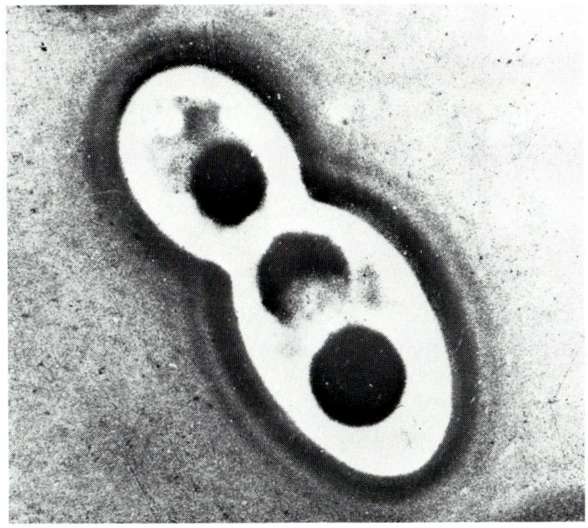

FIGURE 1. Phase contrast micrograph of a budding cell of *Saccharomyces cerevisiae*. (Photograph courtesy of Dr. C. F. Robinow, University of Western Ontario, London, Ontario.)

cerevisiae and many of the subsequent papers in this volume will discuss developments in this field.

A further important "contribution" that yeasts (none of which belong to the genus *Saccharomyces*) make is as pathogens of both plants and animals. In humans, *Candida albicans,* although often present on the skin and mucosal surfaces without causing disease, is an important pathogen.[7] Many other yeasts cause disease in humans or other animals as secondary pathogens or in hosts which lack normal immunity. The conditions under which yeasts induce disease are as varied as the species. They include diabetes, leukemia, various types of cancer, many chronic pulmonary diseases, obesity, pregnancy, laceration of skin surfaces, and a variety of conditions associated with old age.[8]

II. INDUSTRIAL YEAST STRAINS

There is a plethora of literature describing the genetics and biochemistry of laboratory strains of *Saccharomyces cerevisiae,* but there is a general lack of knowledge regarding the genetics and biochemistry of industrial *Saccharomyces* strains. The haploid strain that the molecular biologist employs in the university laboratory as the organism of choice, for reasons already stated in this document, is usually totally unsuitable for use in the industrial world. Brewing yeasts and many other industrial yeasts have been selected over time for exactly those characteristics which render them so unamenable to easy genetic manipulation in the laboratory. They are usually polyploid or aneuploid, lack a mating-type characteristic, sporulate poorly, if at all, and the spores that do form are usually not in fours and exhibit poor spore viability, rendering tetrad analysis difficult.[9] However, in recent studies reported in detail in a later paper in this volume,[10] a marked improvement in the sporogenic ability of a polyploid strain of a brewing strain of *S. cerevisiae* has been observed on cultivation of cells in liquid rather than agar culture media. An analysis of the effects on ascosporogenesis of temperature and of presporulation carbon source (acetate vs. glucose) led to the identification of culture conditions for improved tetrad production. Sporulation at 21°C instead of the usual 27°C gave significant increases in ascus formation. Substitution of glucose with acetate as the presporulation carbon source gave a further increase in total sporulation and a very high induction of tetrads was noted. This increase in sporulation and tetrad formation has permitted the isolation of an array of segregants from the brewing strain in question.

It would appear that the widespread use of polyploid yeasts is no accident. Owing to their multiple gene structure, polyploids are genetically more stable and less susceptible to mutational forces and, thus, can be used continually with a higher degree of confidence than either haploid or diploid strains.

III. FERMENTATION ETHANOL PRODUCTION

In the production of fermentation ethanol, the microorganism being employed should possess a number of important characteristics: (1) rapid and relevant carbohydrate fermentation ability, (2) appropriate flocculation and sedimentation characteristics, (3) genetic stability, (4) osmotolerance (i.e., the ability to ferment concentrated carbohydrate solutions), (5) ethanol tolerance and the ability to produce elevated concentrations of ethanol, (6) high cell viability for repeated recycling, and (7) temperature tolerance.

There are a number of microorganisms that are capable of producing 1% w/v ethanol — *Saccharomyces* spp., *Kluyveromyces fragilis* and *lactis, Candida pseudotropicalis* and *tropicalis, Pachysolen tannophilus, Schwanniomyces alluvius* and *castellii, Endomycopsis fibuligera, Zymomonas mobilis, Clostridium thermocellum* and *thermohydrosulfuricum, Thermoanaerobium brockii*, and *Thermobacteriodes acetoethylicus*. However, at the present time, in excess of 95% of the fermentation ethanol produced globally involves the use of the yeast species *Saccharomyces cerevisiae* and related species.

The requirements of a yeast strain for the production of fermentation ethanol depend very largely on whether the ethanol being produced is for potable or industrial (fuel) purposes. In the production of potable (beverage) ethanol, whether it be for beer, wine, or spirits, the requirements of an acceptable yeast strain can be summarized as follows: (1) the yeast strain should be genetically stable and able to be used over repeated cycles with no genetic change in the culture; (2) it must ferment the medium (wort, must, or mash) in a reasonable period of time to produce 4 to 12% (w/v) ethanol; (3) the fermented medium should be palatable with no off-flavors (i.e., no significant degree of yeast autolysis, phenolic off-flavors, or volatile organo-sulfur compounds); (4) at the end of the fermentation, the yeast should normally be easily removed from the medium (whiskey manufacturing being an exception) either by flocculation or with a centrifuge. This cropped yeast should have a low dead-cell population such that it can be reinoculated into a subsequent fermentation with confidence. By comparison, the requirements of an acceptable yeast strain for industrial ethanol production (solvent and fuel) show similarities and differences to yeast strains employed for potable ethanol production: (1) the yeast strain should be genetically stable with reproducible performance over a number of fermentations, (2) it must be able to ferment the medium in the shortest possible time at a high incubation temperature and produce ethanol near to the theoretical yield value, (3) palatability of fermented product is not a consideration, (4) if the yeast is to be harvested for reuse at the end of the fermentation, it should be in a highly viable state. The ethanol tolerance of the strain is critical. If recycling of the yeast is not required, the yeast plus fermentation medium can be fed directly into the distillation still.

IV. BREWER'S YEAST STRAINS

The two main types of beer are lager and ale. These are fermented with strains of *Saccharomyces uvarum (carlsbergensis)* and *S. cerevisiae*, respectively. Traditionally, lager is produced by bottom-fermenting yeasts at fermentation temperatures between 7 and 15°C. This means that at the end of the fermentation, the yeasts flocculate and collect at the bottom of the fermenter. Top-fermenting yeasts, employed in the production of ales at fermentation temperatures between 18 and 22°C, tend to be less flocculent and loose clumps of cells are carried to the fermenting wort surface adsorbed to carbon dioxide bubbles. Consequently, top yeasts are collected for reuse from the surface of the fermenting wort (a process called skimming), whereas bottom yeasts are collected (or cropped) from the fermenter bottom. The differentiation of lagers and ales on the basis of bottom and top cropping is becoming less and less distinct with the advent of vertical conical bottom fermenters and centrifuges. With centrifuges, nonflocculent yeast strains are required for both lagers and ales, where, as soon as fermentation is complete and before the yeast has had the opportunity to sediment, the fermented medium is passed through a centrifuge in order to separate the yeast from the "green" beer. With conical bottom fermenters, a more sedimentary yeast (lager or ale) settles into the core of the fermenter at the completion of fermentation. It is then removed and a portion of this yeast is reused as an inoculum for a subsequent fermentation.

The requirements of an acceptable brewer's yeast strain can be defined as follows: "In order to achieve a beer of high quality, the yeast cultures must be effective in removing the desired nutrients from the growth medium (i.e., the wort), it must impart the required flavor to the beer, and, finally, the microorganisms themselves must be effectively removed from the fermented wort after they have fulfilled their metabolic role." This definition, although wide ranging and very general, allows the major activities of a brewing yeast strain during the fermentation of wort to be divided into a number of stages: (1) nutrition — sugar and amino acid uptake etc., (2) by-product formation — the excretion of compounds that contribute to the flavor of the product, and (3) removal of the culture from the fermented wort — by flocculation or centrifugation.

In terms of the development and selection of yeast strains for use in a brewery, there are a number of parameters that are important as scale-up criteria: (1) fermentation rate, (2) decrease in specific gravity (°Plato of the wort), (3) taste and flavor match of the final product, (4) fusel oil, ester, and organo-sulfur production, (5) consistent high cell viability, (6) ethanol tolerance and production, and (7) inoculation (pitching rate). More specifically, there are a number of factors that will affect fermentation rate per se, and these include: (1) inoculation (pitching) rate, (2) yeast cell viability, (3) fermentation temperature, (4) wort-dissolved oxygen concentration at pitching, (5) wort-soluble nitrogen concentration, (6) wort fermentable carbohydrate concentration, and (7) yeast storage conditions, e.g., the influence of intracellular glycogen levels.

V. GENETIC MANIPULATION TECHNIQUES

There are a number of methods that are employed in the genetic research and development of brewer's yeast strains. These include hybridization, mutation and selection, rare mating, spheroplast fusion, and transformation. Transformation can be carried out using native DNA, recombinant DNA, or by liposome-mediated DNA transfer. Hybridization cannot be used directly as a means to manipulate brewer's yeast strains for the reasons already discussed in this paper. Nevertheless, hybridization is a technique that has made an invaluable contribution to the field of yeast genetics and is today by no means obsolete. Hybridization has been used in conjunction with more novel genetic techniques to verify the genetic composition of recombinants. It can also be employed to provide a great deal of relevant genetic information about traits that are very germane to brewing fermentation systems. Hybridization has been used to study the genetic control of flocculation, phenolic-off-flavor production, and the uptake of wort sugars and dextrins.

Techniques that have greatest potential and promise as aids in the genetic manipulation of industrial yeast strains are rare mating, mutation and selection, spheroplast or protoplast

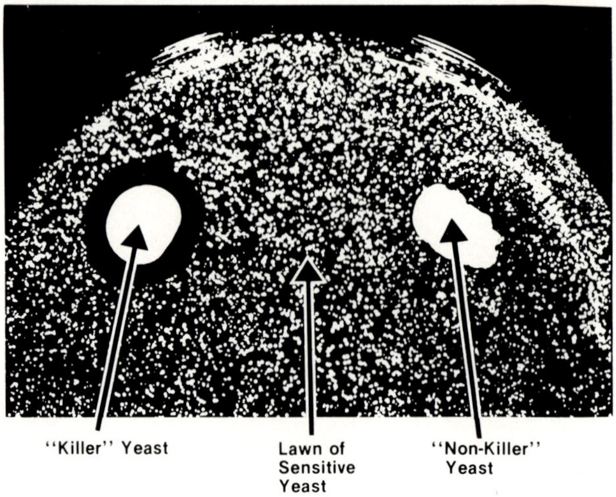

FIGURE 2. *Saccharomyces* brewing yeast with and without zymocidal ("killer") activity.

fusion (also called somatic fusion), and transformation (usually associated with recombinant DNA techniques). All of these methods have a total disregard for ploidy and mating type and thus have great applicability to brewing yeast strains. Rare mating has been used in conjunction with the *kar* (karyogamy defective) strains to introduce zymocidal (killer) activity into brewing strains.[11-13] Mutation and selection have been used to induce auxotrophs and to select easily recognizable characteristics of brewing strains in order that such strains can be employed as spheroplast fusion partners and as recipients for transformation experiments.[14,15] Mutation and selection have also been used to isolate derepressed mutants of brewing strains, such that these strains possess the ability to metabolize maltose in the presence of glucose, and thus have increased wort fermentation rates. Spheroplast fusion has been employed to fuse strains constructed by hybridization with brewing strains in order to introduce the novel capabilities of the hybridized strain into the brewing strain while still maintaining all the characteristics of the latter.[16] Finally, transformation is being employed in this laboratory to introduce genes from non-*Saccharomyces* yeast strains into brewing strains.

VI. RARE MATING AND ZYMOCIDAL ACTIVITY

When nonmating strains are mixed together at a high cell density, a few true hybrids with fused nuclei form which can usually be selectively isolated. An even more useful variation of this technique employs a yeast strain which harbors a specific nuclear gene mutation, designated *kar*. When this strain hybridizes with another strain the nuclei will not fuse, and this permits the formation of cell lines with mixed cytoplasmic contents (heteroplasmons). Rare mating has been employed to transfer "zymocidal" or "killer" factor from laboratory haploid strains to brewing strains. Some strains of *Saccharomyces* spp. (and other yeast genera) secrete a proteinaceous toxin called a "zymocide" or "killer" toxin which is lethal to certain other strains of *Saccharomyces*.[17-20] Toxin-producing strains are termed "killers" and susceptible strains are termed "sensitives". However, there are strains that do not kill and are not themselves killed, and these are called "resistant" (Figure 2).

The "killer" factor has been renamed "zymocide" to indicate that it is only lethal towards yeast and not towards bacteria or cells of higher organisms. Zymocidal yeasts have been recognized to be a serious problem in both batch and continuous fermentation systems.[21,22]

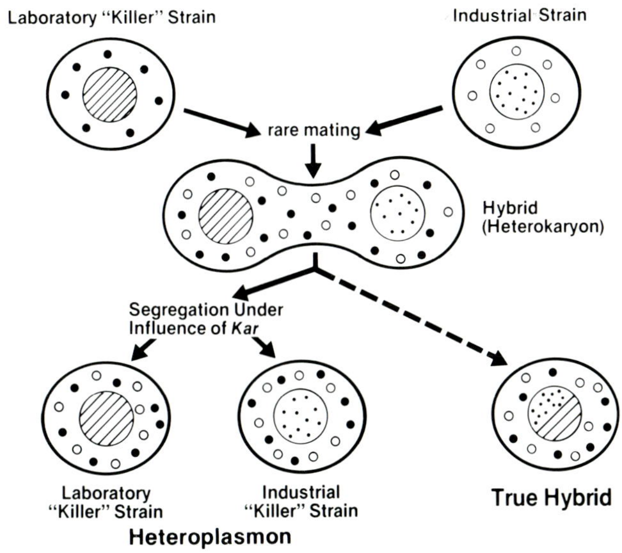

FIGURE 3. Construction of a brewing "killer" strain employing rare mating and the *kar* mutation.

An infection of as little as 1% of the cell population can completely eliminate all the brewing yeast from the fermenter. Thus, the brewer can protect the process from this occurrence in one of two ways: (1) maintain vigorous standards of hygiene to ensure that contamination with a wild yeast possessing zymocidal activity is prevented, or (2) genetically modify the brewery yeast so that it is not susceptible to the zymocidal toxin. The first method is the one most brewers, to date, have been invoking to protect their process; however, genetic manipulation can also be used to produce a brewing strain that is less vulnerable to destruction by a zymocidal yeast infection.

The killer character of *Saccharomyces* spp. is determined by the presence of two species of cytoplasmically located dsRNA.[23-25] M dsRNA (killer plasmid), which is killer strain specific, codes for killer toxin and also for a protein or proteins which render the host immune to the toxin.[26,27] L dsRNA, which is also present in many nonkiller yeast strains, specifies a capsid protein which encapsulates both forms of dsRNA, thereby yielding virus-like particles. Although the killer plasmid is contained within these virus-like particles, the killer genome is not naturally transmitted from cell to cell by any infection process. The killer plasmid behaves as a true cytoplasmic element and requires at least 29 different chromosomal genes (*mak* for its maintenance in the cell). In addition, three other chromosomal genes (*kex1*, *kex2*, and *rex*) are required for toxin production and resistance to toxin.[28] Further studies concerning the killer character in yeasts are reported in a later paper in this volume.[29]

In recent years, the technique of rare mating has been employed to produce hybrids with enhanced fermentation ability.[30,31] A modification of the technique of rare mating, employing the *kar* mutation (karyogamy defective), has been successfully used by a number of research laboratories. The *kar* mutation offers a significant advantage in that it prevents nuclear fusion, and hybrids from such a rare mating can be selected that contain *only* the brewing strain nucleus. However, such hybrids contain the cytoplasm of both parental cells, thereby permitting the introduction of cytoplasmically transmitted characteristics such as killer toxin production into brewing strains without altering the nucleus of the brewing strain. A brewing polyploid lager yeast strain was rare mated with a laboratory "killer" haploid strain and a number of rare mating products isolated (Figure 3).

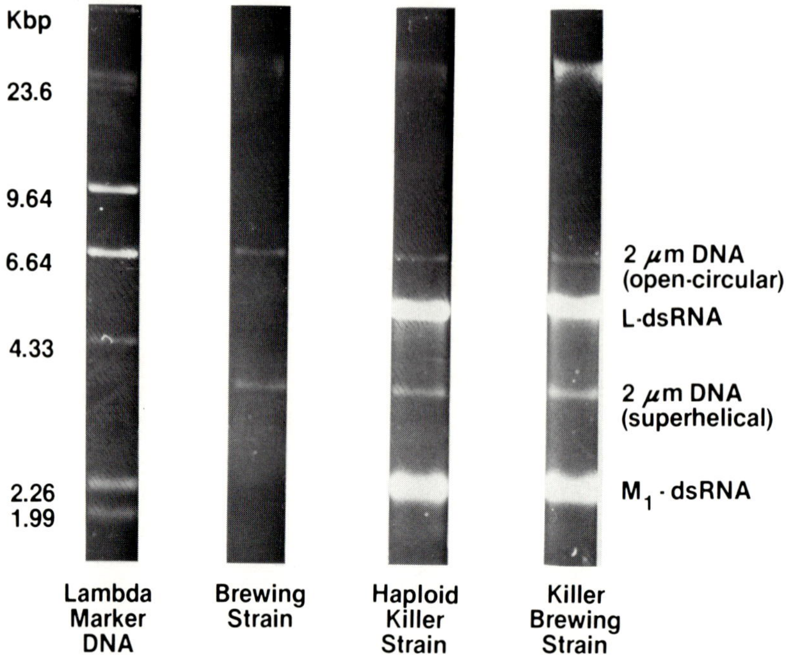

FIGURE 4. Plasmid profile illustrating the dsRNA bands of a ''killer'' yeast.

In addition to biochemical tests to characterize the rare mating products, agarose gel electrophoresis demonstrated that some rare mating products contained not only the 2-μm plasmid (from the parental brewing lager strain), but also the L and M dsRNA plasmids (from the haploid mating partner), which code for killer toxin production (Figure 4).

To determine the effect of the zymocidal lager strain on a typical brewery fermentation, ''killer'' lager yeast was mixed at a concentration of 10% with an ale brewing strain (Figure 5A). The control was the ale strain mixed with 10% ''nonkiller'' lager. The yeast was sampled throughout the fermentation and viable cells determined by plating onto nutrient agar plates incubated at 37°C, a temperature which inhibits the growth of lager yeast, but allows the growth of ale yeast. Within 10 hr the ''killer'' lager strain had almost totally eliminated the ale strain. When the concentration of killer yeast was reduced to 1%, within 24 hr the ale yeast again was eliminated (Figure 5B).

The speed at which this occurs may well make a brewer apprehensive about employing such a yeast in his fermentation cellar, particularly where several yeasts are employed for the production of different beers. An error on an operator's part in keeping lines and yeast tanks separate could result in serious consequences. In a brewery with only one yeast strain this would not be a cause for concern.

An alternative to the ''killer'' strain would be to produce a yeast strain that does not ''kill'', but is ''killer resistant''. That is to say it has received that genetic complement that renders it immune to zymocidal activity. The construction of such a yeast would perhaps be a good compromise, since it does not itself kill, it allays the fear of the brewer that this yeast might kill all other production strains in the plant, and at the same time it is not itself killed by a contaminating yeast with killer activity.

VII. WORT SUGAR UPTAKE

Brewer's wort contains the sugars, sucrose, fructose, glucose, maltose, and maltotriose, together with dextrin material. In the normal situation, brewing yeast strains are capable of

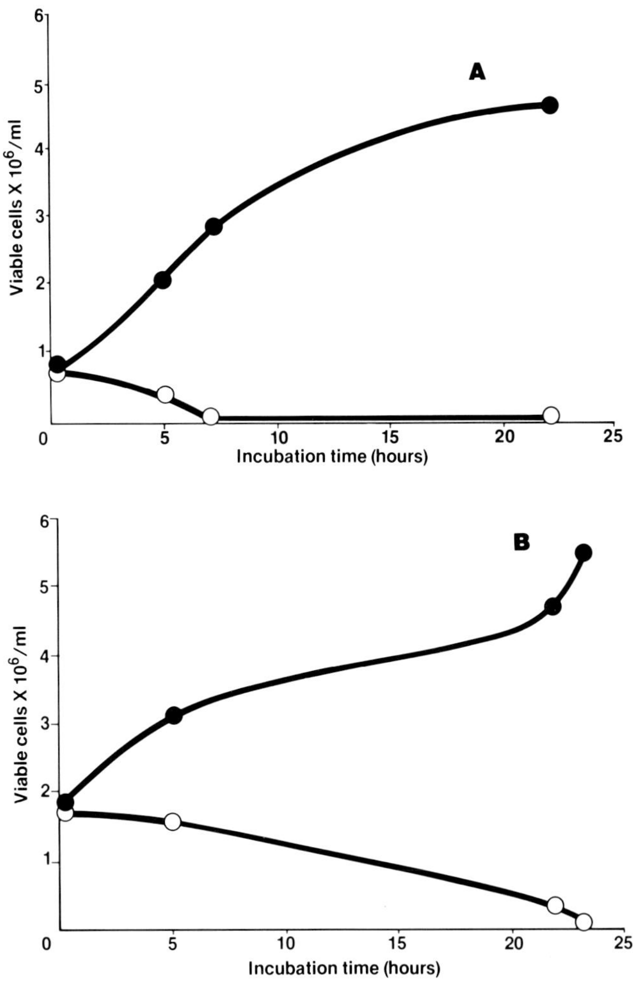

FIGURE 5. (A) Control — growth of ale yeast from mixture of 90% ale yeast plus 10% "nonkiller" lager yeast (●—●). Test — growth of ale yeast from mixture of 90% ale yeast plus 10% "killer" (○—○). (B) Control — growth of ale yeast from a mixture of 99% ale yeast plus 1% "nonkiller" lager yeast (●—●). Test — growth of ale yeast from mixture of 99% ale yeast plus 1% "killer" lager yeast (○—○). The plates were incubated at 37°C which inhibited the growth of both lager yeasts but allowed the growth of the ale yeast.

utilizing sucrose, glucose, fructose, maltose, and maltotriose in this approximate sequence, although some degree of overlap does occur, leaving maltotetraose and the other dextrins unfermented. A major limiting factor in wort fermentation rates is the repression of maltose and maltotriose uptake by glucose.[32] Only when approximately 50% of the wort glucose has been taken up by the yeast will the uptake of maltose commence. In other words, in most strains of *Saccharomyces cerevisiae* and related species maltose utilization is subject to control by carbon catabolite repression, such that even in the presence of maltose, the maltose-utilizing system will be inactivated by high concentrations of glucose. In order to illustrate this phenomenon, two brewing strains of *S. cerevisiae* (coded 154 and 3001) have been studied. In both instances (Figures 6 and 7), the presence of glucose repressed the uptake of maltose. In an attempt to isolate spontaneous mutants of these strains in which the presence

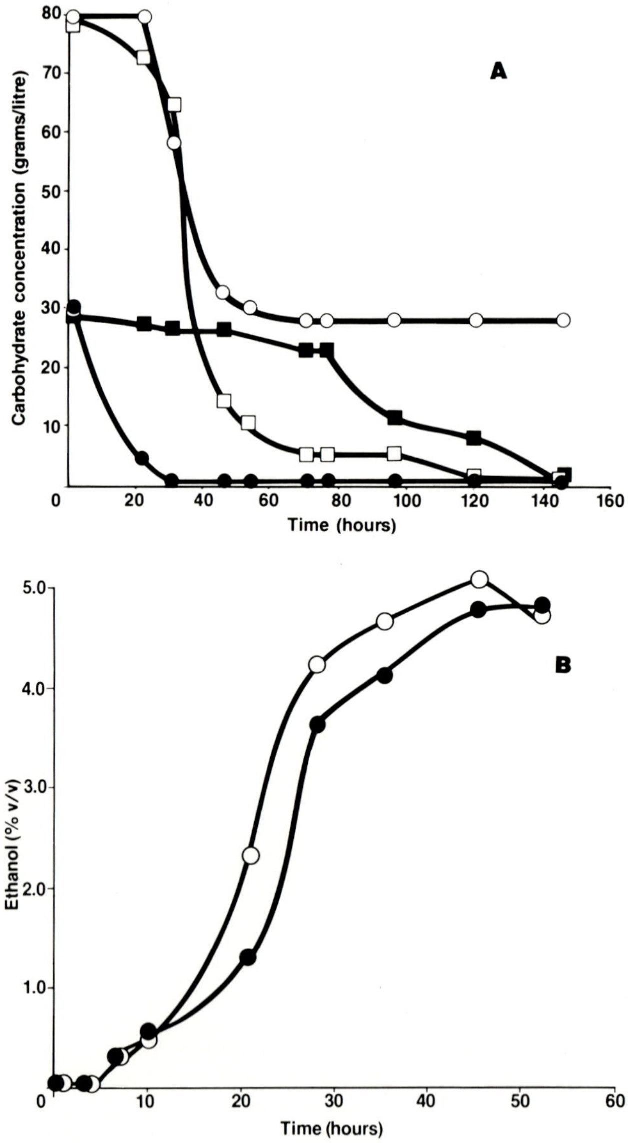

FIGURE 6. Fermentation characteristics of a brewing strain of *Saccharomyces cerevisiae* strain 154 and its derepressed maltose 2-DOG mutant. (A) Carbohydrate uptake in an 8% (w/v) maltose 3% (w/v) glucose-peptone-yeast extract medium. Glucose uptake: parent (●—●) and mutant (■—■); maltose uptake: parent (○—○) and mutant (□—□). (B) Ethanol production in a 12°P wort: parent (●—●) and mutant (○—○).

of glucose in the medium did not repress maltose uptake, mutants capable of growth in the nonmetabolizable glucose analog 2-deoxyglucose (2-DOG) were isolated.

A number of stable 2-DOG mutants were found to be capable of utilizing maltose in the presence of significant concentrations of glucose. For example, 2-DOG-resistant mutants of *S. cerevisiae* (strain 154) in a synthetic medium containing both maltose (8% [w/v] and glucose (3% [w/v]) were able to completely metabolize the maltose (Figure 6A), whereas in the same medium, maltose uptake was slow with the parental strain and only 60% complete when fermentation ceased. Fermentation and ethanol formation rates in 12°P wort were also increased in the 2-DOG mutants when compared to the parental strain (Figure 6B).

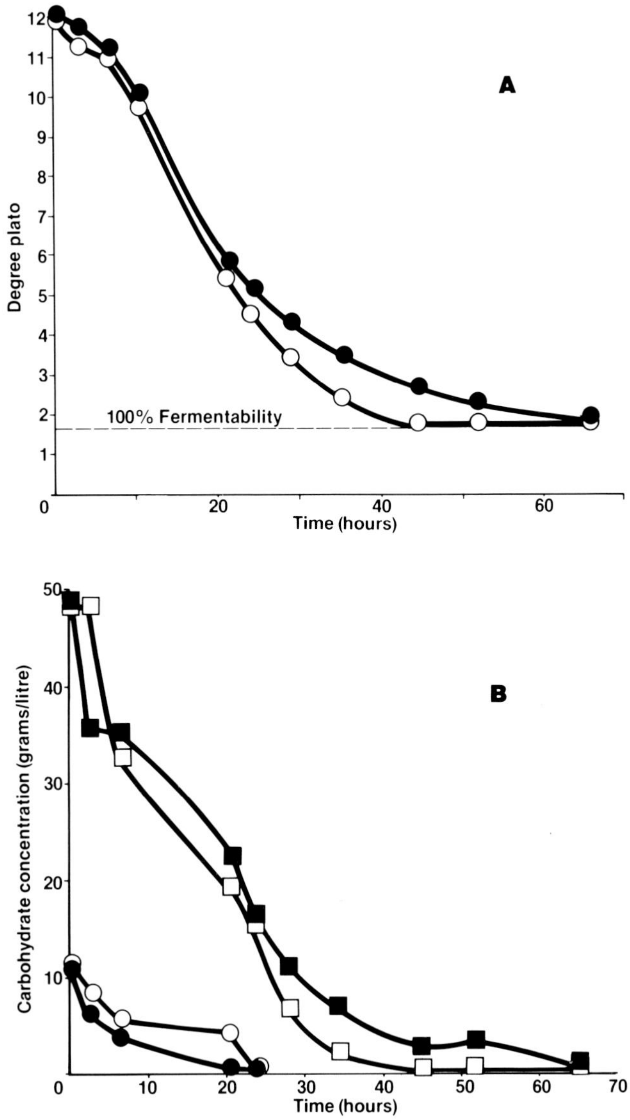

FIGURE 7. Fermentation characteristics of a brewing strain of *Saccharomyces cerevisiae* (strain 3001) and its derepressed maltose 2-DOG mutant in a 12°P wort. (A) Fermentation rate: parent (●—●) and mutant (○—○). (B) Glucose uptake: parent (●—●) and mutant (○—○); maltose uptake: parent (■—■) and mutant (□—□).

Similar results were obtained with the other strain of *S. cerevisiae* (strain 3001) studied; maltose uptake in 2-DOG mutants was not repressed by glucose. In 12°P wort, the overall fermentation rate was significantly faster in the 2-DOG mutants (Figure 7A), with complete fermentation being achieved in 45 hr compared to 65 hr in the parental strain. This increased fermentation rate was due to an increased maltose uptake rate in the 2-DOG mutants and compared to the parent strain, with glucose having little influence upon maltose uptake in the mutants (Figure 7B). A trained taste panel operating in the triangular mode determined that the beer produced with 2-DOG mutants of both strains (i.e., 154 and 3001) was not significantly different from that produced using the parental yeast strain; all beers were

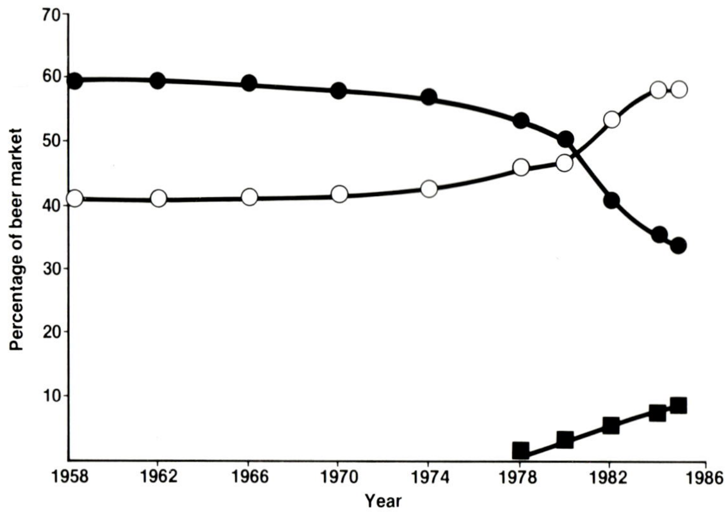

FIGURE 8. Consumption of ale (●—●), lager (○—○), and light (■—■) beer in Canada.

produced under similar pilot plant brewing conditions. The mechanism by which 2-DOG-resistant mutants are derepressed is far from understood, but such mutants have been reported to possess diminished levels of hexokinase.[33] Possibly, hexokinase is associated with a general regulatory system in yeast, involving overall repression.

Bailey and Woodward[34] have described a mutant allele (designated *grr-1*) in *S. cerevisiae* that is characterized not only by 2-DOG resistance, but insensitivity to glucose repression for invertase, maltase, and galactokinase, as well as the mitochondrial enzyme cytochrome *c* oxidase. In this case the levels of hexokinase activity towards both fructose and glucose were approximately threefold higher in mutant clones. The molecular mechanism by which catabolite repression is effected in yeast is still unknown, although many theories have been discussed. Nevertheless, derepressed mutants of industrial yeast strains will be of significant economic importance, particularly with respect to increased control of fermentation rate.

VIII. LOW CARBOHYDRATE (LOW CALORIE) "LITE" BEER

Low carbohydrate (usually low calorie) beer in North America represents a significant share of the beer volume; it constitutes 13% of the market, or 22 million hectoliters (581 million gallons) per annum in the U.S. and 8% or 1.7 million hectoliters (45 million gallons) per annum in Canada (Figure 8).

Several procedures for producing low carbohydrate beers are being employed or contemplated; most of them center on one of the following techniques: (1) dilution of regular-strength beer with water, (2) addition of fungal α-amylase or glucoamylase and bacterial pullulanase to the wort during fermentation, (3) use of glucose, fructose, and sucrose as an adjunct, (4) use of brewing yeast strains with amylolytic activity, and (5) use of a malt enzyme preparation during mashing or fermentation.

As previously discussed, typical brewing yeast strains are capable of utilizing sucrose, glucose, fructose, maltose, and maltotriose, leaving the maltotetraose and the larger dextrins unfermented. However, yeasts of the species *Saccharomyces diastaticus*[35] have been classified as a distinct species from that of *S. cerevisiae* due to the fact that the former produces the extracellular enzyme glucoamylase, also called amyloglucosidase, (α-1,4 glucan glucohydrolase, E.C.3.2.1.3.). This enzyme possesses the ability to cleave α-1,4 and, in some cases, α-1,6 linkages, releasing glucose from the nonreducing end of starch chains.

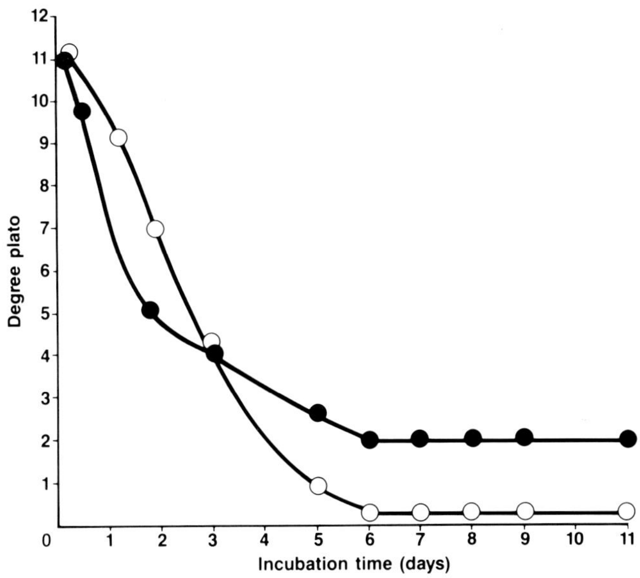

FIGURE 9. Static fermentation of a brewer's wort (40-ℓ scale) by a *Saccharomyces cerevisiae* brewing strain *(dex)* (●—●) and a *S. diastaticus* diploid *(DEX1/DEX1, DEX2/DEX2, STA3/STA3)* (○—○).

On the other hand, α-amylase (α-1,4 glucan 4 glucanohydrolase, E.C.3.2.1.1) is an enzyme which hydrolyzes random α-1,4 glucosidic linkages by-passing α-1,6 bonds. Three genes have been identified that are associated with glucoamylase production of *S. diastaticus* — *DEX1, DEX2,* and *STA3.*[35] Using classical hybridization techniques, a diploid strain containing the *DEX* and *STA* genes in the homozygous condition has been constructed and its fermentation rate studied in brewer's wort under static fermentation conditions (Figure 9). The initial fermentation rate of this strain was slower than a production ale-brewing strain, however, the *DEX*-containing strain fermented the wort to a greater extent than the brewing strain due to the partial hydrolysis of the dextrins by the action of glucoamylase. Thus, *S. diastaticus* strains possess the capacity to produce beer which has been fermented to a high degree and is desirable in the production of low-carbohydrate beer; however, it should be noted that beer produced by these strains has a characteristic phenolic off-flavor. Phenolic off-flavors in beer are due on many occasions to the presence of 4-vinyl guaiacol (4-VG) which arises by the enzymatic decarboxylation of ferulic acid, a wort constituent.[36] It has been found by Tubb et al.[12] that a single, dominant, nuclear gene designated *POF* (Phenolic Off-Flavor) codes for the ferulic acid decarboxylation enzyme. Therefore, strains possessing the *POF* gene can produce the enzyme capable of decarboxylating ferulic acid. Whereas brewing *Saccharomyces* strains normally cannot decarboxylate ferulic acid, all the *S. diastaticus* strains initially studied produced 4-VG in the presence of ferulic acid. Assuming that the *POF* and *DEX* genes are independent characteristics, it should be possible to construct a strain containing the *DEX* but not the *POF* gene by means of hybridization. Thus, a haploid that was *DEX* positive and that carried the *POF* characteristic was mated with a dextrin-negative phenolic-off-flavor-negative haploid. The resultant diploid fermented dextrin and decarboxylated ferulic acid. When tetrad dissection was carried out a 2:2 segregation for dextrin fermentation and a 2:2 segregation for phenolic off-flavor was obtained. The *DEX* and *POF* genes segregated independently of each other, therefore, it was possible to select haploids that were *DEX* positive and *POF* negative. Subsequently, a diploid with the genotype *DEX2/DEX2, pofo/pofo* was constructed and a fermentation of an 11.3°P wort was conducted

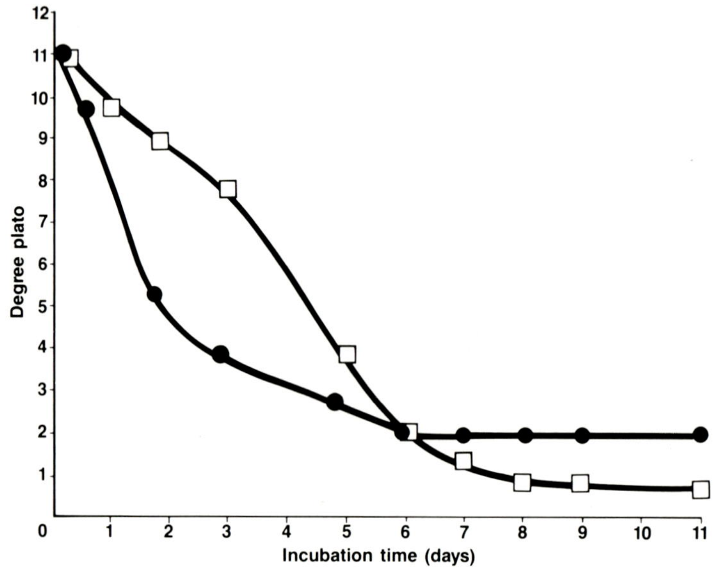

FIGURE 10. Fermentation characteristics of a polyploid *Saccharomyces cerevisiae* strain (●—●) and a diploid *DEX2/DEX2, pof/pof, S. diastaticus* strain (□—□) in a brewer's wort. Static fermentation at 21°C, 0.3% (w/v) pitching rate.

(Figure 10). Although the initial wort attenuation rate was found to be slower than that of a polyploid ale yeast strain, the yeast was capable of superattenuating the wort, i.e., it was able to hydrolyze part of the dextrins into glucose, which is readily fermentable, whereas the brewing strain was unable to utilize the dextrins. ''Expert'' taste panel assessment has deemed the beer produced from this dextrin-positive diploid to be rather winey and to have a slightly sulfury character, however, the characteristic phenolic off-flavor associated with the *POF* gene (4-VG) could not be detected.

The production of glucoamylase by strains of *S. diastaticus* is subject to carbon catabolite repression by glucose and other sugars.[37] Subjecting two such strains to the 2-DOG selection technique, stable, spontaneous mutants derepressed for the formation of glucoamylase and mutants derepressed for maltose utilization have been isolated. In 12°P wort with a 2-DOG starch mutant of *S. diastaticus* (strain 1393), the level of glucoamylase was increased fivefold when compared to the parental strain (Figure 11A). The overall fermentation rates of the starch and maltose mutants were increased when compared to the parental strain (Figure 11B).

A second strain of *S. diastaticus* (strain 1400) gave similar results to strain 1393. In 12°P wort, the 2-DOG maltose mutants of this strain possessed increased maltose uptake rates together with an overall faster fermentation rate when compared to the parental strain (Figures 12A and 12B).

A further problem with the use of strains of *S. diastaticus* (*POF* or *pof*) for the production of beer is the fact that the extracellular glucoamylase is heat stable and remains active in the packaged beer after pasteurization. As a consequence, upon storage at 21°C an increasing concentration of glucose can be found in the beer (Figure 13). Current activities are directed to the development of brewer's yeast strains that are able to secrete a thermolabile glucoamylase.

IX. DIASTATIC YEASTS FOR DISTILLED ETHANOL PRODUCTION

The fermentation of starches to ethanol by yeasts requires pretreatment of the substrate in order to produce fermentable sugars. This pretreatment consists of three steps: gelatini-

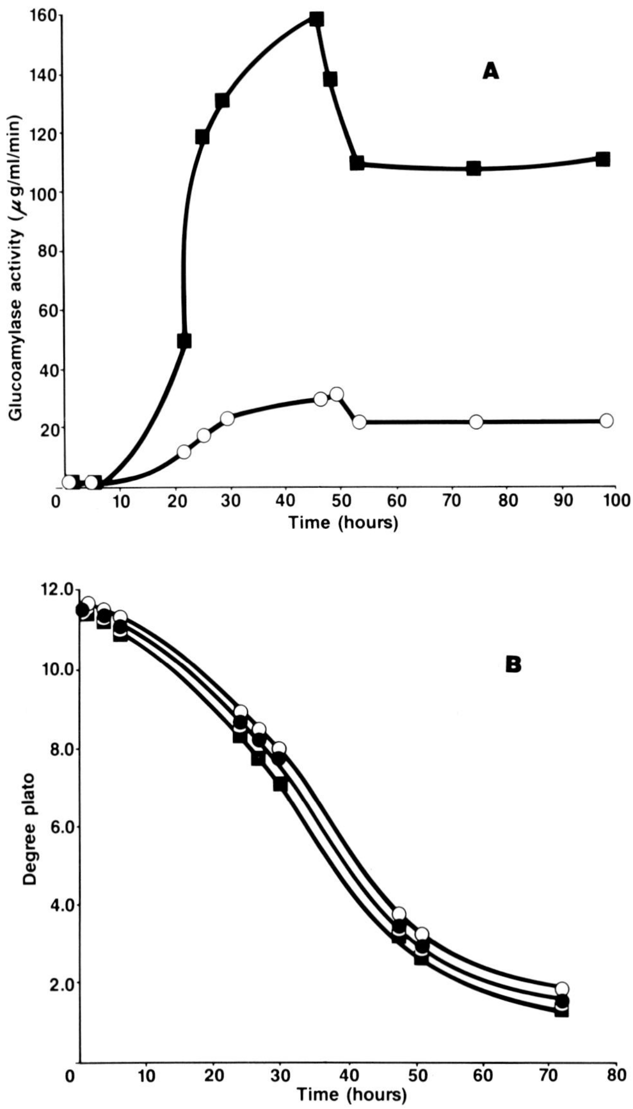

FIGURE 11. Fermentation characteristics of a diploid strain of *Saccharomyces diastaticus (DEX1/DEX1, DEX2/DEX2, STA3/STA3, MAL6/mal6)* (strain 1393) and its derepressed maltose and starch 2-DOG mutants in a 12°P wort. (A) Production of glucoamylase: parent (○—○) and derepressed starch mutant (■—■); (B) fermentation rate: parent (○—○), derepressed maltose mutant (●—●), and derepressed starch mutant (■—■).

zation, liquefaction, and saccharification. Gelatinization requires heat and free water and must precede liquefaction. Liquefaction, the dispersion of starch molecules into an aqueous solution, is accomplished by the use of heat and amylolytic enzymes. Heat-stable bacterial α-amylases or malt enzymes may be employed. During liquefaction starch molecules are only partially hydrolyzed, producing a form of carbohydrate which cannot be assimilated by ethanol-producing yeasts such as *Saccharomyces cerevisiae*. Therefore, the partially hydrolyzed starch molecules must be converted to lower molecular weight sugars, such as glucose and maltose, by a process known as saccharification. This may be accomplished

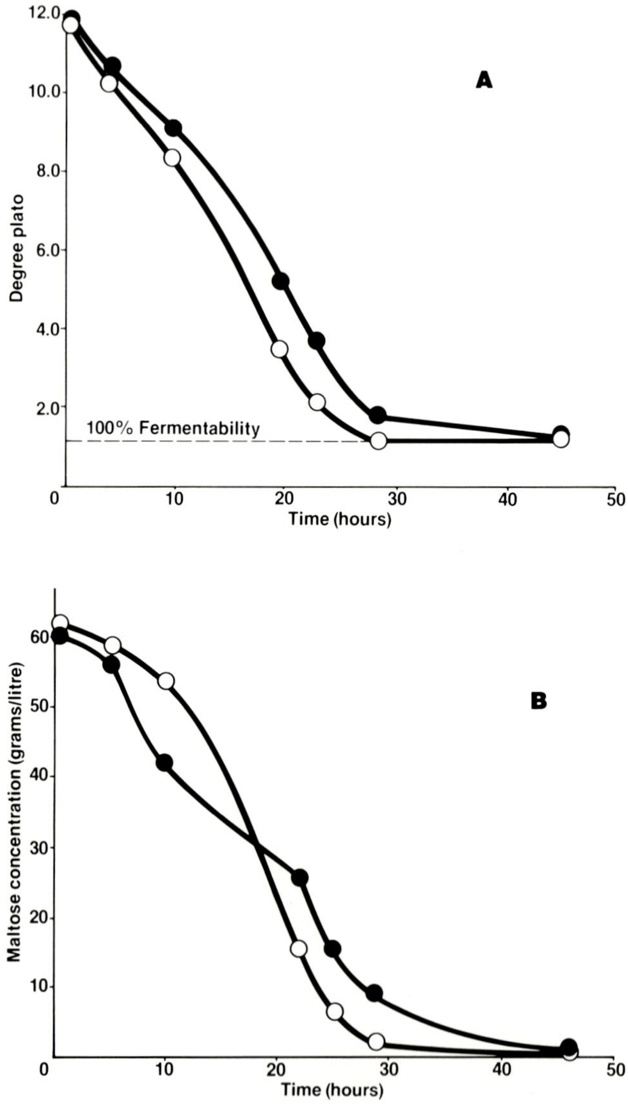

FIGURE 12. Fermentation characteristics of a strain of *Saccharomyces diastaticus* (strain 1400) and its derepressed maltose 2-DOG mutant in a 12°P wort. (A) Fermentation rate: parent (●—●) and mutant (○—○); (B) maltose uptake: parent (●—●) and mutant (○—○).

enzymatically, usually by the addition of fungal glucoamylases to the fermentation vessel at the time of inoculation. The saccharifying glucoamylases represent a significant fraction of the total cost of producing ethanol. Reduction of the amount of added glucoamylase could significantly decrease the cost of the final product.

It has been found possible to decrease glucoamylase addition to starch mash fermentations by employing yeasts which actively produce and secrete glucoamylase, e.g., strains of *S. diastaticus*.[38,39] A genetically manipulated diploid strain of *S. diastaticus* (strain 1393, *DEX1/DEX1, DEX2/DEX2, STA3/STA3, MAL6/mal6*) was studied as the glucoamylase-producing strain and compared to a strain of *S. cerevisiae* (strain 254) which was unable to produce and secrete glucoamylase. The fermentation performance of the two strains was compared in a corn mash with and without added glucoamylase. When the *S. diastaticus* (strain 1393)

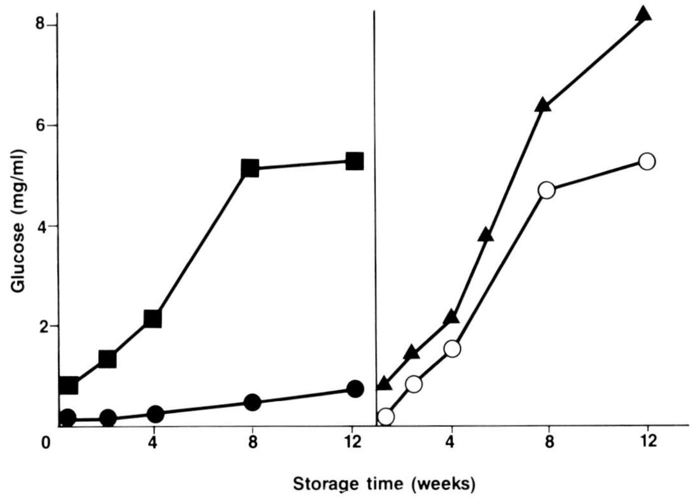

FIGURE 13. Production of glucose in pasteurized beer during storage at 21°C by a polyploid *Saccharomyces diastaticus DEX* strain (■—■), an ale brewing *dex* strain (●—●), a haploid *DEX1, DEX2, STA3* strain (▲—▲), and a haploid *DEX2* strain (○—○).

culture was employed, added glucoamylase concentrations could be significantly decreased without reducing ethanol production or sugar uptake (Figures 14 and 15). Reduction of the added glucoamylase concentration from 0.1% by volume, based upon the substrate weight, to 0.05% resulted in no significant decrease in ethanol yield or sugar uptake, the reduction of added glucoamylase being possible because the hybrid yeast is able to produce and secrete its own glucoamylase.

Production of glucoamylase by the *S. diastaticus* strain and subsequent reduction of added glucoamylase represent a potential financial savings for the ethanol producer. Since the cost of glucoamylase may exceed $3.75 (U.S.) per liter, a 50% reduction of this enzyme can result in a significant decrease in cost. For example, an ethanol manufacturer that is producing 100 million liters of 95% ethanol per year uses 190 million kilograms of starch (assuming that the fermentation of 20% (w/v) starch yields 10% (v/v) ethanol). For 190 million kilograms of starch, 0.19 million liters of glucoamylase are required (based upon an addition of 0.1 ℓ of glucoamylase per 100 kg of starch). this results in an annual enzyme cost of approximately $700,000 U.S. assuming an enzyme cost of $3.75 U.S. per liter. Reduction of the added glucoamylase by 50% represents an annual saving of approximately $350,000 U.S.

X. CONCLUSIONS

Although yeast is the oldest of the biotechnology microorganisms, it is still a very important microorganism that is being employed in the modern biotechnology revolution. Current research is being directed into two areas, the first being to improve its capacity and efficiency to produce ethanol and carbon dioxide from a wide variety of substrates, and the second is to employ yeast to synthesize and excrete a wide variety of proteins or peptides that possess therapeutic and commercial potential. The wide-ranging papers being published in this volume illustrate the spectrum of activities currently ongoing in both academic and industrial research laboratories.

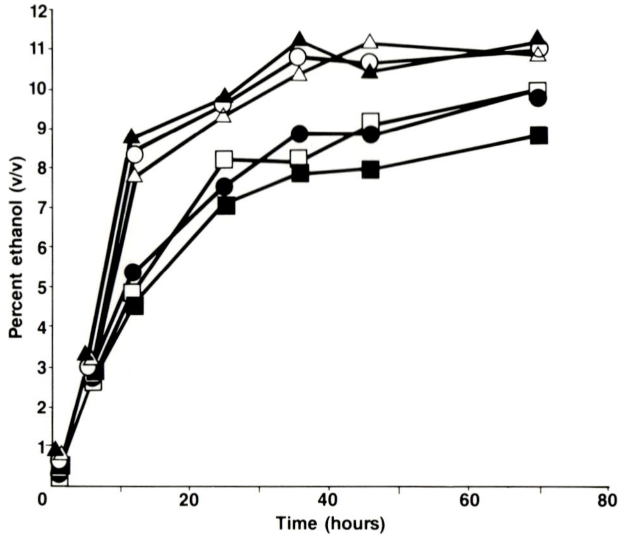

FIGURE 14. Ethanol production by hybrid strain 1393. 0.0% v/w AMG (■—■), 0.025% v/w AMG (□—□), 0.04% v/w AMG (●—●), 0.05% v/w AMG (△—△), 0.075% v/w AMG (○—○), 0.10% v/w AMG (▲—▲).

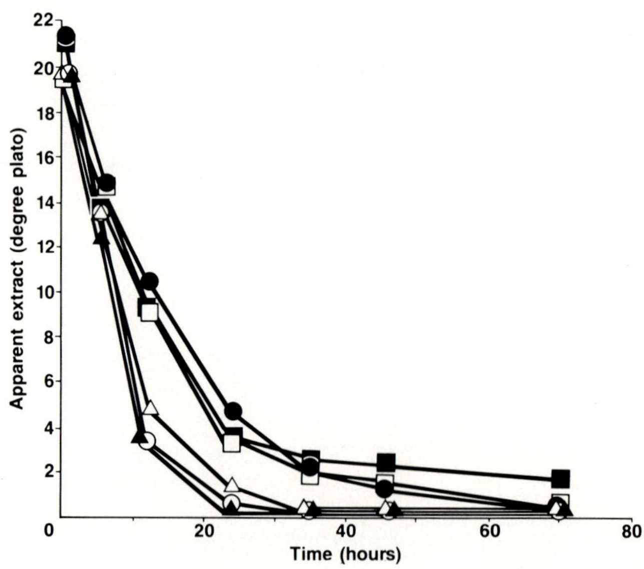

FIGURE 15. Apparent extract values for hybrid strain 1393. 0.0% v/w AMG (■—■), 0.025% v/w AMG (□—□), 0.04% v/w AMG (●—●), 0.05% v/w AMG (△—△), 0.075% v/w AMG (○—○), 0.10% v/w AMG (▲—▲).

ACKNOWLEDGMENT

The authors are grateful to their colleagues, C. Bilinski and C. Panchal, for help and advice during the preparation of this manuscript. Part of the research being discussed was funded by a PILP Contract from the National Research Council of Canada, which is gratefully acknowledged.

REFERENCES

1. **Kreger-van Rij, N. J. W.,** *The Yeasts — a Taxonomic Study*, 3rd ed., Elsevier, Amsterdam, 1984.
2. **Upshall, A. and McKnight, G. L.,** The development of filamentous fungi as expression hosts, in *Biological Research on Industrial Yeasts*, Vol. 1, Stewart, G. G., Russell, I., Klein, R. D., and Hiebsch, R. R., Eds., CRC Press, Boca Raton, Fla., 1987, chap. 10.
3. **Hitzman, R. A., Leung, D. W., Perry, L. J., Kohr, J. W., Levine, H. L., and Goeddel, D. W.,** Secretion of human interferons by yeast, *Science*, 219, 620, 1983.
4. **Molzahn, S. W.,** Fermentation Processes and Their Products, European Patent Appl. 0,147,198 A2, 1985.
5. **Smith, R. A., Duncan, M. J., and Moir, D. T.,** Heterologous protein secretion from yeast, *Science*, 229, 1219, 1985.
6. **Mortimer, R. K. and Schild, D.,** Genetic map of *Saccharomyces cerevisiae*, edition 9, *Microbiol. Rev.*, 49, 181, 1985.
7. **Shepherd, M. G., Poulter, R. T. M., and Sullivan, P. A.,** *Candida albicans:* biology, genetics, and pathogenicity, *Annu. Rev. Microbiol.*, 39, 579, 1985.
8. **Odds, F.,** *Candida and Candidosis*, Leicester University Press, Leicester, U.K., 1979.
9. **Stewart, G. G.,** The genetic manipulation of industrial yeast strains, *Can. J. Microbiol.*, 27, 973, 1981.
10. **Bilinski, C. A., Hatfield, D. E., Sobczak, J. A., Russell, I., and Stewart, G. G.,** Analysis of sporulation and segregation in a polyploid brewing strain of *Saccharomyces cerevisiae*, in *Biological Research on Industrial Yeasts*, Vol. 2, Stewart, G. G., Russell, I., Klein, R. D., and Hiebsch, R. R., CRC Press, Boca Raton, Fla., 1987, chap. 4.
11. **Hammond, J. R. M. and Eckersley, K. W.,** Fermentation properties of brewing yeast with killer character, *J. Inst. Brewing*, 90, 167, 1984.
12. **Tubb, R. S., Searle, B. A., Gooday, A. R., and Brown, A. J. P.,** Rare mating and transformation for construction of novel brewing yeasts, in *Proc. 18th Int. Eur. Brew. Conv. Congr., Copenhagen*, IRL, London, 1981, 487.
13. **Young, T. W.,** Brewing yeast with anti-contaminant properties, in *Proc. 19th Int. Eur. Brew. Conv. Congr., London*, IRL, London, 1983, 129.
14. **Bilinski, C. A., Sills, A. M., and Stewart, G. G.,** Morphological and genetic effects of benomyl on polyploid brewing yeast: induction of auxotrophic mutants, *Appl. Environ. Microbiol.*, 48, 813, 1984.
15. **Delgado, M. A. and Conde, J.,** Benomyl as a tool for the parasexual genetic analysis of brewing yeast, in *Proc. 19th Int. Eur. Brew. Conv. Congr., London*, IRL, London, 1983, 465.
16. **Stewart, G. G., Murray, C. R., Panchal, C. J., Russell, I., and Sills, A. M.,** The selection and modification of brewer's yeast strains, *Food Microbiol.*, 1, 289, 1984.
17. **Philliskirk, G. and Young, T. W.,** The occurrence of killer character in yeast of various genera, *Antonie van Leeuwenhoek J. Microbiol. Serol.*, 41, 147, 1975.
18. **Rogers, D. and Bevan, E. A.,** Group classification of killer yeasts based on cross-reactions between strains of different species and origin, *J. Gen. Microbiol.*, 105, 199, 1978.
19. **Woods, D. R. and Bevan, E. A.,** Studies on the nature of the killer factor produced by *Saccharomyces cerevisiae*, *J. Gen. Microbiol.*, 51, 115, 1968.
20. **Young, T. W. and Yagiu, M.,** A comparison of the killer character in different yeasts and its classification, *Antonie van Leeuwenhoek J. Microbiol. Serol.*, 44, 59, 1978.
21. **Maule, A. P. and Thomas, P. D.,** Strains of yeast lethal to brewery yeasts, *J. Inst. Brewing*, 79, 137, 1973.
22. **Taylor, R. and Kirsop, B.,** Occurrence of a killer strain of *Saccharomyces cerevisiae* in a batch fermentation plant, *J. Inst. Brewing*, 85, 325, 1979.
23. **Bevan, E. A. and Somers, J. M.,** Somatic segregation of the killer (k) and neutral (n) cytoplasmic genetic determinants in yeast, *Genet. Res.*, 14, 71, 1969.
24. **Fink, G. R. and Styles, C. A.,** Curing of a killer factor in *Saccharomyces cerevisiae*, *Proc. Natl. Acad. Sci. U.S.A.*, 69, 2846, 1972.
25. **Gunge, N.,** Yeast DNA plasmids, *Annu. Rev. Microbiol.*, 37, 253, 1983.
26. **Berry, E. A. and Bevan, E. A.,** A new species of double stranded RNA from yeast, *Nature (London)*, 239, 279, 1972.
27. **Bevan, E. A., Herring, A. J., and Mitchell, D. J.,** Preliminary characterization of two species of dsRNA in yeast and their relationship to the "killer" character, *Nature (London)*, 245, 81, 1973.
28. **Wickner, R. B.,** The killer double-stranded RNA plasmids of yeast, *Plasmids*, 2, 303, 1979.
29. **Thomas, D. Y., Vernet, T., Boone, C., Greene, D., Lolle, S., and Bussey, H.,** The yeast killer toxin expression systems functional analysis and biotechnological applications, in *Biological Research on Industrial Yeasts*, Vol. 1, Stewart, G. G., Russell, I., Klein, R. D., and Hiebsch, R. R., Eds., CRC Press, Boca Raton, Fla., 1987, chap. 13.

30. **Stewart, G. G., Bilinski, C. A., Panchal, C. J., Russell, I., and Sills, A. M.,** The genetic manipulation of brewer's yeast strains, in *Microbiology 1985*, ASM Publications, Washington, D.C., 1985, 367.
31. **Russell, I., Jones, R., and Stewart, G. G.,** The influence of cytoplasmic control on the expression of nuclear genes in brewing yeast strains, in *Proc. 20th Int. Eur. Brew. Conv. Congr., Helsinki,* IRL, London, 1985, 235.
32. **Stewart, G. G., Russell, I., and Sills, A. M.,** Factors that control the utilization of wort carbohydrates by yeast, *MBAA Tech. Q.*, 20, 1, 1983.
33. **Entian, K. D., Zimmerman, F. K., and Scheel, I.,** A partial defect in carbon catabolite repression in mutants of *Saccharomyces cerevisiae* with reduced hexose phosphorylation, *Mol. Gen. Genet.*, 156, 99, 1977.
34. **Bailey, R. B. and Woodward, A.,** Isolation and characterization of a pleiotropic glucose repression resistant mutant of *Saccharomyces cerevisiae, Mol. Gen. Genet.*, 193, 507, 1984.
35. **Erratt, J. A. and Stewart, G. G.,** Genetic and biochemical studies on glucoamylase from *Saccharomyces diastaticus,* in *Current Developments in Yeast Research,* Stewart, G. G. and Russell, I., Eds., Pergamon Press, Toronto, 1981, 177.
36. **Wackerbauer, K., Kossa, T., and Tressl, R.,** Die bildung von phenolen durch hefen, in *Proc. 16th Int. Eur. Brew. Conv. Congr., Amsterdam,* IRL, London, 1977, 495.
37. **Sills, A. M., Sauder, M. E., and Stewart, G. G.,** Amylase activity in certain yeasts and fungal species, *Dev. Ind. Microbiol.*, 24, 295, 1983.
38. **Whitney, G. K., Murray, C. R., Russell, I., and Stewart, G. G.,** Potential cost savings for fuel ethanol production by employing a novel hybrid yeast strain, *Biotechnol. Lett.*, 7, 349, 1985.
39. **Stewart, G. G., Russell, I., and Panchal, C. J.,** Genetically Stable Allopolyploid Somatic Fusion Product Useful in the Production of Fuel Alcohol, Canadian Patent 1,119,593, 1986.

Chapter 2

APPLICATIONS OF ULTRAHIGH CELL DENSITY YEAST FERMENTATIONS

Donald O. Hitzman, G. H. Wegner, and L. K. Shay

A highly efficient process to produce large quantities of yeast was developed by Phillips Petroleum Company in response to a world need for protein. Early work centered on the use of gas-oil fractions or *n*-paraffins as substrates for yeast growth. However, such processes using water-immiscible substrates with their inherent limitations led to the discovery at Phillips that alcohols, and particularly methanol, would be the preferred substrate. Such oxygenated products of hydrocarbon gases eliminated the problems of substrate solubility and carry-over of potential carcinogenic residues. Thus, methanol became the feedstock of choice to produce large quantities of yeast for single-cell protein applications.

The use of alcohols required the development of novel approaches to grow cells at concentrations of alcohols that were considered biocidal. This led to the development of a continuous fermentation mode in which methanol, although fed continuously at rates of up to 40% of the feed volume, remained at essentially a zero level in the fermentor. Thus, a fermentation process for growing yeasts in a continuous substrate-limiting condition was developed.

Although this process of growing yeast on methanol was extremely successful, both in regard to yield and growth rate, the productivity of the system remained relatively low. Since the yield and growth rate were not amenable to significant improvement, the best way to improve productivity was to increase the cell concentration. At that time cell densities of over 40 g/ℓ dry weight of yeast, when grown on a continuous basis, would be considered satisfactory. However, if cell densities could be increased to 80 or 120 g/ℓ, while retaining the yield and dilution rate, the productivity would also be doubled or tripled. Fermentations at such high cell densities, while maintaining optimum growth conditions such as temperature control and adequate O_2 transfer, would require an improvement in fermentor design. Thus, it was necessary to develop high efficiency oxygen and heat transfer fermentors concurrently with medium manipulation in order to achieve the high productivities needed for an economically successful process technology. The approach of simultaneously balancing yield, growth rate, feed concentrations, nutrient requirements, oxygen and heat transfer capabilities, etc. was pursued. In addition, an understanding of the interrelationship between these various fermentation parameters and the yeast metabolism was necessary to maintain such operations. At the same time, constraints were always present, since the process had to be practical for engineering scale-up and economic reasons. This led to studies on cell recovery, nutrition, and composition aspects of the product development, enzyme isolation, animal feeding, and toxicological testing. Thus, an entire technology was developed for the continuous growth of yeasts at cell concentrations of over 120 g/ℓ dry cell weight. This technology was termed an Ultrahigh Density Direct Dry Process and is diagrammed in Figure 1. The system uses special fermentors designed to optimize oxygen and heat transfer. Oxygen transfer rates in excess of 800 mmol $O_2/\ell \cdot hr$ are achieved in the process systems. Biofluid side heat transfer coefficients in excess of 1100 $J/M^2 \cdot S \cdot K$ can be obtained with specially designed tube baffle heat exchangers. In system operations, the medium composition and fermentation conditions for the chosen yeast and feedstock are tailored to give maximum yield of cells and maximum protein content. The selected substrate is used as the sole carbon and energy source and is the limiting nutrient throughout the fermentation. The concentration of dissolved oxygen is maintained at as low a level as possible while still having no oxygen limitation. The air flow rate is maintained at a level at which there is no CO_2 inhibition. Ammonia is supplied

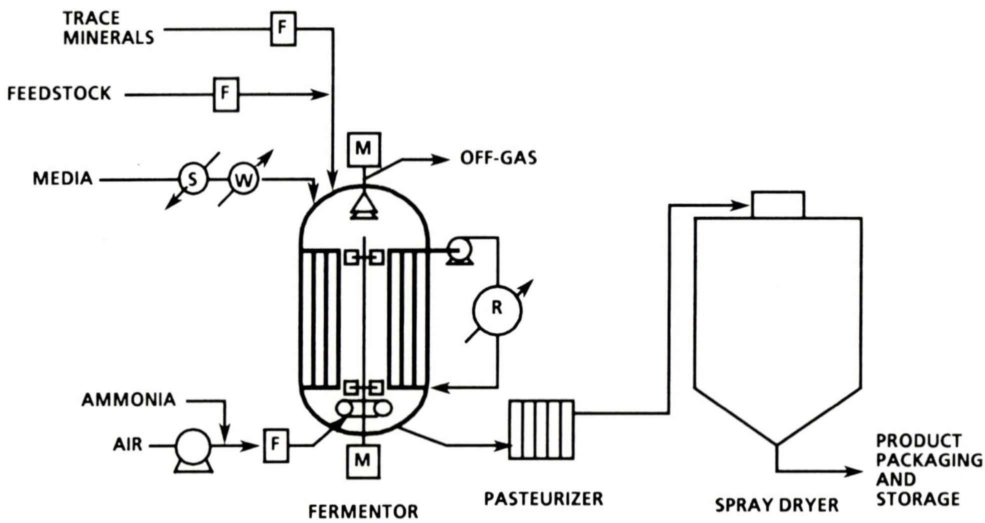

FIGURE 1. Phillips continuous Provesteen® single-cell protein process.

for pH control and as a nitrogen source. The system is controlled at the optimum growth temperature for the selected culture. Under this regime of continuous optimized growth conditions, the culture always remains in the most active log state of growth and maximum productivity is maintained.

The fermentor effluent, having a dry product solid content of above 120 g/ℓ, is dried directly after heat killing the yeast. Complex operations, such as filtration and centrifugation, are eliminated by this process, as are difficult problems, such as maintaining sanitary conditions in such equipment. The elimination of major waste streams and of the need to recycle spent medium simplifies plant design, engineering, and operations.

This technology was expanded to include feedstocks other than methanol and to include such diverse substrates as glucose, sucrose, molasses, ethanol, whey, etc. In addition, the number and types of microorganisms which can be grown at such high cell densities in a continuous mode have been expanded to include many species of yeast and bacteria such as *Escherichia coli, Bacillus,* and fungi such as *Aspergillus.* This significantly increases the versatility of this process technology and offers many opportunities for the practical production of cells and products which previously were uneconomical or difficult to achieve. Representative examples of the potential of this technology to grow a yeast at ultrahigh densities on a variety of substrates are shown in Table 1. The fermentation parameters for the growth of *torula* yeast (*Candida utilis*) on various substrates utilizing this technology are shown in Table 2. The data include results obtained using small laboratory-size fermentors, as well as pilot plant operations in a 1500-ℓ-size unit. Test data with other substrates and yeasts show results similar to those reported.

It is evident that this development of a process to grow cells on a continuous basis at ultrahigh cell densities has immediate applications for a wide range of processes (Table 3). The system, when viewed as a technology that continuously delivers cell concentrations almost equivalent to that obtained after a centrifugation step, offers all the possibilities that a centrifugation preparation offers. In addition, it has the advantage of the cells remaining in an uncontaminated log state of growth. This means such cells are available for secondary fermentations or enzymatic conversions. The process will also deliver genetically modified cells that have the potential to initiate selective product formation.

As an example of applications of this technology, one of the immediate targets involved the production of enzymes. In this connection, the ability of specific yeasts, such as *Pichia*

Table 1
REPRESENTATIVE CELL MASS YIELDS OF *CANDIDA UTILIS* STRAIN NRRL Y1082 GROWN AT ULTRAHIGH CELL DENSITIES

Substrate	Retention time (hr)	Cell mass (g/ℓ)	Yield (%)	Productivity g/ℓ · hr
Ethanol	5.0	127	81	25.4
Sucrose	4.7	160	52	34
Molasses	5.9	163	51	27.6
Glucose molasses[a]	6.4	117	50	18.3
Purified molasses[b]	3.6	136	51	37.8

[a] Glucose molasses is the waste product of the conversion of molasses to fructose.
[b] Purified molasses is a demineralized product of molasses.

Table 2
FERMENTATION PARAMETERS OF ULTRAHIGH DENSITY FERMENTATIONS FOR PRODUCTION OF *TORULA* YEAST

Parameter	Sucrose pilot plant	Ethanol laboratory	Molasses Laboratory	Molasses Pilot plant
Dilution rate (hr^{-1})	0.16	0.2	0.17	0.20
pH	3.9	3.8	4.5	4.5
Temperature (°C)	32—35	32	35	35
Fermentation broth				
Total dry solids (g/ℓ)	150	140	320	220
Washed cells (g/ℓ)	135	127.4	160	105
Productivity (g/ℓ/hr)				
Based on total dry solids	25	28	55.9	44.1
Based on washed cells	23	25.4	27.1	21
Yield (gram cells per gram substrate)	0.52	0.8		
Oxygen transfer (mmol O_2/ℓ/hr)	780	714	1100	870
Heat transfer (kcal/ℓ/hr)	92	70	126	100

or *Hansenula,* to utilize methanol as the sole carbon source results in the production of large quantities of alcohol oxidase. This enzyme concentration is far in excess of the enzymatic amounts needed for the metabolic processing of methanol. In addition, the alcohol oxidase is packaged in solid, crystalline form in peroxisomes. The reason for this is not known, but it has been a boon for the efficient production of the enzyme. In addition, other enzymes in the elucidated pathway of methanol metabolism in *Pichia* are catalase, formaldehyde dehydrogenase, *S*-formyl glutathione hydrolase, and formate dehydrogenase. This technology permits the isolation of such enzymes to be considered as feasible. The same type of overproduction of enzymes, such as lactase, can be achieved by simply changing the substrate and yeast.

Table 3
APPLICATIONS OF ULTRAHIGH CELL DENSITY CONTINUOUS FERMENTATION TECHNOLOGY

Large-scale production of cells
Utilization of fermentable waste streams
Preparation of enzymes
Growth of rDNA cultures
Delineation of cell nutritional requirements
Production of metabolic products
Seed fermentor operations
Secondary downstream fermentations
Alternative to cell immobilization
Optimization of fermentation parameters
Culture improvements

Another alternative in applying this technology is to use the high cell density effluent stream from the fermentor to feed a secondary reactor which is maintained under a different condition. Thus, it is possible to grow the cells on one substrate and pass these cells to a second fermentor in which a different substrate, or nonmetabolizable compound, is present. Such a reactor could be operated under a completely different set of conditions and may even involve conditions that are unfavorable for growth, yet which would be conducive for enzymatic conversions of product formation. This ability to modify conditions of a continuous stream of cells at high density offers unique opportunities for product production.

As can be envisioned, such a system is well suited to the production of genetically engineered cultures. For example, the gene which encodes alcohol oxidase in *Pichia* has been cloned in view of the high per-cell level of the alcohol oxidase (>30% of soluble protein). The gene has been characterized by sequence analysis, the sites of transcription and translation initiation defined, and the promoter and regulatory sequences identified. A β-galactosidase gene fusion with the 5′ end of the alcohol oxidase gene has been constructed and the heterologous production of the foreign protein β-galactosidase achieved. These studies were extended to demonstrate the practical fermentor production of other rDNA products.

Another application of this technology should be cited, because it is so important in developing and optimizing any cell growth system. This involves the delineation of nutritional and metabolic requirements of a culture. When cells are grown in the log state at these concentrations, the effect of adding any stimulating or inhibitory material is almost instantly reflected in the oxygen uptake, as measured by the dissolved oxygen content of the fermentor. Thus, by operating in a steady-state condition at a constant dissolved oxygen level, the addition of the material will cause either a positive (greater oxygen uptake) or negative (decreased oxygen uptake) change in the dissolved oxygen level. By monitoring such shifts in dissolved oxygen, the nutritional requirements of a culture can be accurately defined very quickly. This is a powerful technique to identify factors which affect cell growth.

Although the system as described involves no centrifugation step following fermentation, in many cases such a step would be desirable to remove the supernatant for product recovery or to wash the cells. Such washed cells could then be readily immobilized or otherwise treated. Other techniques, such as cell disruption to free product or enzymes, are successful in preparing cell homogenates for further processing.

Thus, this ultrahigh cell density technology offers a variety of opportunities and has great potential in the area of industrial yeast biochemistry and molecular biology. The list of applications of this fermentation technology is constantly being expanded, and it can be expected that each investigator will use and develop this technology to his own advantage.

Chapter 3

MOLECULAR TAXONOMY OF INDUSTRIAL YEASTS

Cletus P. Kurtzman

TABLE OF CONTENTS

I. INTRODUCTION

Yeasts are emerging as the dominant microorganisms for use in biotechnology, and with this prominence comes the need for a more complete understanding of both the metabolic potential and the genetics of these simple eukaryotes. Genetic modifications to exploit metabolic potential may be achieved through mutagenesis, traditional genetic crosses, and cell fusion, or by the introduction of novel genes via molecular biological techniques such as transformation.

Perhaps the major constraint on the successful application of any of these procedures is the genetic distance between gene donor and recipient. This has long been evident in genetic crosses instituted by matings and is becoming apparent as new combinations are constructed from cell fusions.[1] It also appears true when relying upon newer techniques such as transformation. While shuttle vectors with wide host ranges are well known, the presence of unique plasmids in many species[2,3] tells us that there are indeed limitations, if not on genetic transfer, then on replication and expression.

As a consequence, it becomes useful to know the genetic distance between microorganisms whose properties one may wish to combine and to have expressed. Ideally, one might gauge this genetic distance from a knowledge of taxonomy. Practically, this has not been true, because most schemes of classification are constructed from phenotypic characters of unknown phylogenetic significance. The use of molecular comparisons offers a remedy to this problem. For example, three recent publications by the same group of authors described the same fermentation, but by a "different yeast", two of which were in one genus while the third was in another genus. Had these authors investigated the taxonomy of the species, they would have learned that DNA relatedness studies published 4 years earlier showed all three taxa to be conspecific. One other unfortunate aspect of this work was the implication that this particular fermentation reaction was common to a wide variety of yeasts.

II. TAXONOMIC IMPACT OF NUCLEAR DNA COMPARISONS

A. Nucleic Acid Isolation and Purification

The methodology for isolation and purification of yeast DNA has been dealt with in detail elsewhere, and the following account will serve only as a brief review to direct attention to some of the more pertinent references. For many yeasts, DNA isolation is relatively straightforward. Cell wall breakage may be effected by mechanical or lytic means. Mechanical methods have the advantage of being relatively rapid, but unless cell breakage is carefully monitored, considerable shearing of the DNA may result. The Braun cell homogenizer, one of the more frequently used mechanical devices, causes breakage by its shaking of a buffered cell suspension in the presence of 0.45-mm glass beads.[4-6] An inexpensive variation of this technique involves placing the buffered cell and bead mixture into a test tube and "vortexing" the mixture.[7,8] Less frequently reported methods include grinding acetone-dried cells in a ball mill[9] and grinding cells in liquid nitrogen with a mortar and pestle.[10,11] Sonication is seldom used for cell breakage because of the danger of extreme DNA shearing.

Lytic methods offer the possibility of isolating DNA with a higher molecular weight than might be obtained by mechanical cell disruption. Meyer and Phaff[12] and Bicknell and Douglas[13] used a modification of the Smith and Halvorson[14] technique in which yeast cells undergo autolysis in the presence of saline EDTA, 2-mercaptoethanol, and sodium lauryl sulfate. The occasional failure of this method has prompted the widespread use of mechanical disruption. The enzyme preparations Glusulase and Zymolyase are used to prepare osmotically fragile spheroplasts which may be disrupted to give high-molecular-weight DNA. Cryer et al.[15] have given details of these procedures. Zymolyase generally promotes more rapid cell wall dissolution than Glusulase, but neither enzyme preparation is particularly effective with basidiomycetes. Brady,[16] however, reported the dissolution of *Rhodotorula* cell walls by an enzyme preparation from culture filtrates of *Penicillium lilacinum*, and Rhodes and Kwon-Chung[17] produced protoplasts from blastospores of *Cryptococcus neoformans* with the multienzyme preparation mutanase.

Following rupture or lysis of the cells, DNA isolation and purification may be carried out by the methods of Marmur[18] and Bernardi et al.[19] as detailed by Price et al.[5] Basically, this consists of making the cell suspension 1 *M* in sodium perchlorate and 1% in sarcosate, and emulsifying with an equal volume of chloroform-isoamyl alcohol (24:1, vol/vol). After incubation and phase separation by centrifugation, DNA is precipitated from the aqueous phase with cold ethanol. The DNA is treated with α-amylase, pancreatic RNase, and pronase. Following enzyme removal, the DNA is again ethanol precipitated, spooled, redissolved, and further treated with α-amylase, pancreatic RNase, and T_1 RNase. Additional purification can include hydroxylapatite fractionation and banding in a cesium chloride gradient.

Some yeasts have active DNases that escape denaturation during the usual extraction procedures and cause sufficient degradation of the DNA to prevent spooling. Several methods have produced varying degrees of success in the deactivation of nucleases. Diethyl pyrocarbonate used at 0.5 to 1.0% at the time of cell breakage gave encouraging results in studies of *Lipomyces* spp.[20] and *Sporidiobolus*.[21] Kurtzman and co-workers[22] found the spermine-spermidine-sucrose buffer of Morris, as cited by Timberlake,[23] to more effectively inactivate nucleases in certain species than the sucrose buffer frequently used in yeast studies.[5]

Unless care is taken to isolate intact nuclei and mitochondria, the preceding methods will give a mixture of DNA from these two organelles. Because of its lower guanine + cytosine content, mitochondrial DNA may be separated from nuclear DNA through preparative CsCl density centrifugation. Methods for assessing DNA purity include chemical analyses, as well as spectrophotometric measurements, and have been summarized by Price et al.[5] and Johnson.[24]

Techniques for isolation of RNA have been presented by Gillespie and Spiegelman,[25] Bicknell and Douglas,[13] Kirby,[26] and Johnson.[24] These references also include procedures for RNA-DNA reassociation experiments. Although the cellular RNA concentration is manyfold higher than that of DNA, the nearly ubiquitous occurrence of RNases and their marked stability make isolation of undegraded RNA more difficult than the isolation of high-molecular-weight DNA.

B. DNA Base Composition and Its Taxonomic Uses

Nuclear DNA base composition, expressed as molar percentages of guanine + cytosine (G + C), is the simplest of the molecular techniques that allow resolution of taxa. Base composition may be determined from thermal denaturation profiles (melts), buoyant density in cesium salt gradients generated by ultracentrifugation, chemical analysis, absorbance ratios, or high-pressure liquid chromatography of nucleotides or free bases.[24] Most investigators use thermal denaturation,[27] but Bak[28] showed that results obtained by this method can be greatly affected by sample impurities and/or minor DNA species and must therefore be interpreted with caution. Cesium chloride buoyant density determinations[29] generally show the greatest accuracy, since they are unbiased by the presence of contaminating RNA, mitochondrial DNA, and other impurities, such as carbohydrates and proteins.

The taxonomic uses of G + C values are mainly exclusionary, because yeast species range in G + C content from approximately 28 to 70 mol% and overlap between unrelated species is inevitable. For example, *Debaryomyces hansenii* and *D. marama* have a G + C content of 39.0 and 39.1 mol%, respectively, yet they show only 8.4% base sequence complementarity.[5] Consequently, because similar G + C content does not necessarily indicate conspecificity, we must ask how dissimilar G + C values need to be in order to state with relative certainty that two strains represent different species. The answer to this question depends to some extent on the method used for determination of base composition. When buoyant density is used, the span of G + C values among strains of a species is about 1%,[5,6,30] which is approximately coincident with the accuracy of the method, i.e., a standard deviation of 0.00 to 0.50 mol%. Thermal denaturation shows greater variation, and strains of a species may have a G + C range of as much as 2%.[31] Thus, strains showing a difference in G + C contents in excess of 1.0 to 1.5% by buoyant density or 2.0 to 2.5% by thermal denaturation may be expected to be different species. Among the yeasts it is possible to predict the taxonomic subdivision from G + C content. Ascomycetous species range from 28 to 52%, whereas basidiomycetous species range from 49 to 70%.[32] Species within a genus frequently have G + C values that lie within a range of 10%.

C. DNA Relatedness

The methods for assessing DNA relatedness vary, but short of actual sequencing, rely on measuring the extent and stability of renatured DNA strands from the test pair, i.e., the fidelity of complementary base pairing. Depending on the method, the DNA may or may not need to be labeled with radioisotopes. Because of technical difficulties, heavy isotope labels are seldom used.

In vivo labeling of DNA is frequently done using ^{14}C, ^{3}H, or ^{32}P. Each isotope has its particular advantages and disadvantages. The DNA may also be labeled in vitro with ^{125}I[33-35] or with ^{14}C, ^{3}H, or ^{32}P by nick translation.[36,37] DNA labeled with ^{125}I in vitro must be highly purified, which requires banding in CsCl in a preparative ultracentrifuge. Following labeling, DNA reassociation reactions may be carried out using the membrane method, which involves immobilizing single strands of one of the DNA species onto nitrocellulose filters[38] or other appropriate support and allowing sheared single strands of the second DNA to react with the immobilized DNA. Alternatively, both DNA species may be allowed to react in free solution, and the extent of reassociation may be assessed by percent binding of the resulting duplexes to hydroxylapatite[39] or resistance to hydrolysis by S_1 nuclease.[40] Price et al.[5] and Johnson[24] provide references and detailed procedures for these methods.

In addition to following DNA reassociation with isotopically labeled preparations, the reactions may also be monitored spectrophotometrically by measuring the kinetics of duplex formation. This technique has been refined to the point where results are comparable to data obtained with isotopes.[6,41-43] Relatedness may also be estimated from the thermal stability of the renatured heterologous DNA as compared with renatured homologous DNA. This may be done spectrophotometrically by monitoring thermal melts or with labeled DNA by thermal elution from hydroxylapatite columns.[44] Brenner et al.[45] equated 1° of difference in thermal stability with 1% mismatch in base pairs. Lower thermal stability usually correlates well with lower DNA homology.

A number of factors affect DNA renaturation reactions and so may have a marked effect on the results obtained. Highly purified DNA preparations are essential. Contaminating protein or RNA can result in falsely high or low relatedness values, depending on whether the contaminants act as single- or double-stranded DNA in the assay system.[39,46]

Several classes of DNA occur in eukaryotes, and their presence affects estimates of relatedness between taxa because it is the intent of most studies to measure genomic similarity from complementarity of single-copy nuclear DNA sequences. Christiansen et al.[47] reported up to 15% rapidly renaturing sequences in the nuclear DNA (nDNA) of yeasts, which may be interpreted as analogous to the repeated sequences found in higher eukaryotes.[48] However, the presence of rapidly renaturing, plasmidlike 2-μm circular DNA in *Saccharomyces cerevisiae*,[49] with a buoyant density similar to nDNA, suggests the proportion of repeated sequences in nDNA to be smaller than first thought. Similar circular DNAs have been reported from *Pichia canadensis (Hansenula wingei)* and *Candida (Torulopsis) glabrata*.[50,51] Linear DNA plasmids have been discovered in *Kluyveromyces lactis*.[52]

Nuclear DNA represents the preponderance of cellular DNA in yeasts and has a kinetic complexity of 6.5 to 14 × 10^9 daltons or about two to five times that found in *Escherichia coli*.[47,53] Between 5 and 20% of the total DNA in *S. cerevisiae* is mitochondrial in origin. It is arranged in closed circles with a G + C content noticeably lower than the nDNA.[54] Sedimentation studies[55] and measurements of contour length[53,56] suggest a molecular mass of 46 × 10^6 daltons. The mitochondrial DNA (mtDNA) contour length of several other yeasts has been reported to be 25 to 50% that of *S. cerevisiae*.[50,51]

If comparisons of taxa are to be made from single-copy nDNA, mtDNA, plasmidlike circular DNA, and repeat sequence nDNA must be removed. Circular plasmidlike DNA as well as circular mtDNA may be relatively easily separated from nDNA by centrifugation in cesium chloride in the presence of ethidium bromide. Williamson and Fennell[57] emphasized that resolution is poor if mtDNA should break into linear fragments during isolation, a not uncommon happening. However, they point out that linear mtDNA may be effectively resolved in cesium chloride gradients when reacted with 4′,6-diamidino-2-phenylindole (DAPI), a fluorescent dye that preferentially binds with AT-rich DNA. Repeated sequences may be removed from nDNA by partial reassociation of sheared DNA fragments followed by fractionation with hydroxylapatite. Since the rate of reassociation is affected by concentration, multiple copies duplex first and can be removed because of their ability to bind to hydroxylapatite.[48] Price et al.[5] used this method to remove repeated sequences from yeast DNA for taxonomic studies. The fractionation, undoubtedly, also removes mtDNA and plasmidlike DNA, because both occur as multiple copies, and their small molecular weight would allow rapid reassociation when in significant concentration. The same effect may be obtained in spectrophotometric determinations of relatedness if the time between melting and cooling to the reassociation temperature (T_{m-25}) is not too rapid. Relatedness values of strains common to the work of Price et al.[5] and of Kurtzman et al.[6] were essentially identical, showing that similar results can be obtained with rather different methodologies.

Optimal DNA reassociation occurs at 25 to 30°C below the midpoint of the melting curve (T_m), and this is dependent on base composition, since GC base pairs have greater thermal stability than do AT pairs.[27,58] If reassociation occurs at a temperature higher than optimum, any mismatched sequences are less likely to pair, whereas temperatures under the optimum may allow the mismatch to appear complementary. Consequently, we must view DNA relatedness values obtained from whole genome comparisons as an estimate of overall base sequence similarity under defined conditions, rather than as a certain percentage of the DNA that has precisely the same base sequence.

D. Extent of Resolution Correlated with Genetic Hybridization

Guidelines establishing the extent of nuclear DNA relatedness exhibited between strains of microbial species have resulted primarily from studies of bacteria. On the basis of extensive comparisons between species of enteric bacteria, Brenner[39] proposed that strains showing 70% or greater DNA relatedness should be considered members of the same species. Johnson[59] compared numerous anaerobic bacteria and arrived at essentially the same conclusions. Among the yeasts, Price et al.[5] examined species in four genera and proposed that strains having 80% or greater DNA relatedness were conspecific, an opinion expressed earlier by Martini and Phaff[9] from their studies in *Kluyveromyces*. However, the guidelines proposed in all of the studies were based on strains that had been characterized by traditional morphological and physiological tests. Another approach would be to compare the fertility of sexually reproducing species with extent of DNA relatedness in order to correlate DNA complementarity with actual biological phenomena.

Van der Walt and Johannsen[60] and Johannsen[61] approached this problem through hybridization studies in the essentially homothallic genus *Kluyveromyces*, where they were able to compare their results with DNA relatedness data obtained earlier by Martini[62] and Martini and Phaff.[9] Van der Walt and Johannsen mated auxotrophic yeast strains, attained by *N*-

Table 1
CORRELATION OF MATING REACTION AND DNA COMPLEMENTARITY AMONG CLOSELY RELATED HETEROTHALLIC ASCOMYCETOUS AND BASIDIOMYCETOUS YEASTS

Species	Mating reaction	DNA relatedness (%)	Ref.
Filobasidiella neoformans × *F. bacillispora*	Fair conjugation 0—30% Basidiospore viability (F_1 progeny; F_2 not determined)	55—63	137
Issatchenkia scutulata var. *scutulata* × *I. scutulata* var. *exigua*	Good conjugation Ascospores viable: F_1 = 5%; F_2 = 17%	21—26	30
Pichia amylophila × *P. mississippiensis*	Good conjugation Ascospores not viable	20—27	6
Hansenula bimundalis × *H. americana*	Poor conjugation Ascospores not produced	21	111
H. alni × *H. wingei*	Poor conjugation Ascospores not produced	6	110, 138
I. orientalis × *I. occidentalis*	Infrequent conjugation Ascospores not produced	3—8	30

methyl-*N*′-nitro-*N*-nitrosoguanidine mutagenesis, on media that supported growth of prototrophic recombinants. Utilizing this method, they observed various combinations of prototrophic colonies at a frequency of 10^{-2} to 10^{-8}. Taxa with high DNA relatedness showed a higher frequency of recombination, but in a few cases, taxa having low DNA complementarity also gave a considerable number of recombinants. Lack of crosses through successive generations of progeny has made interpretation of these data difficult.

Kurtzman and co-workers[6,30] were able to correlate genetic relatedness and DNA complementarity by working with heterothallic strains, thereby circumventing problems inherent in the *Kluyveromyces* study where auxotrophic mutants of homothallic strains were employed. In addition, Kurtzman et al.[6,30] isolated progeny from crosses by micromanipulation. In the first of these studies, *Pichia amylophila* and *P. mississippiensis* were shown to mate well, but ascospores were not viable. The two species showed only about 25% DNA relatedness, thus correlating lack of fertility with low DNA complementarity. However, the relationship between *Issatchenkia scutulata* var. *scutulata* and *I. scutulata* var. *exigua* was somewhat different. These two taxa also showed only about 25% DNA relatedness, but genetic crosses gave 3 to 6% viable ascospores. Furthermore, crosses between these F_1 progeny gave 17% viability among F_2 progeny. Reciprocal crosses among the F_1 progeny were fertile, as were backcrosses to the parents. From this, it appears that all stocks had essentially homologous chromosomes and that progeny were neither amphidiploids nor aneuploids. Clearly then, the lower limit of DNA-DNA homology values suggesting species delimitation is not yet well defined, but base sequence divergence, as estimated from whole genome comparisons, may be as great as 75% before genetic exchange can no longer occur. Data in Table 1 summarize the preceding findings and provide further comparisons of other taxa. In these studies, progressively less DNA relatedness parallels decreasing mating competence and fertility. These data also indicate that resolution goes no further than the detection of sibling species. Beyond this, all species, regardless of extent of kinship, show essentially less than 10% complementarity.

Knowledge of strain histories, when coupled with DNA relatedness measurements and information on mating reactions, allows some insight into the process of species formation. For example, *Issatchenkia scutulata* var. *scutulata* was isolated in Hawaii while *I. scutulata* var. *exigua* was found in the Pacific Northwest.[63] Genetic crosses suggest both varieties to

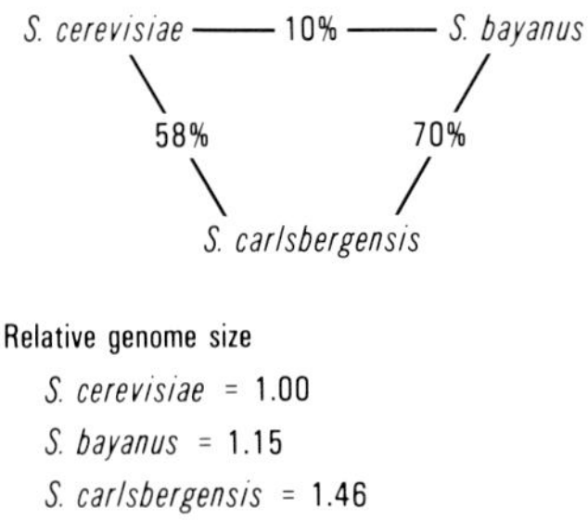

FIGURE 1. Extent of nuclear DNA relatedness between type strains of *Saccharomyces cerevisiae*, *S. bayanus*, and *S. carlsbergensis*. Genome sizes were estimated from reassociation kinetics and *S. cerevisiae* was assigned a relative size of 1.00 for this comparison. (After Martini, A. V. and Kurtzman, C. P., *Int. J. Syst. Bacteriol.*, 35, 508, 1985.)

have essentially homologous chromosomes, and we can speculate that divergence resulted from base sequence substitution owing to geographical separation.[30] This is probably also true for *Pichia (Hansenula) bimundalis* and *P. (H.) americana*. The former species is indigenous to Europe and Japan, while the latter is known only from the western U.S. By contrast, *P. mississippiensis* and *P. amylophila* were found in the frass of adjacent pine trees in Mississippi.[6] At the time of isolation, strains of *P. mississippiensis* formed haploid ascospores, but *P. amylophila* gave only diploid ascospores. Repeated single-spore selections eventually yielded an isolate that produced tetrads on sporulation. It seems likely that *P. amylophila* was initially a strain of *P. mississippiensis* that failed to undergo meiosis at the time of ascosporulation. Since the incipient new species was competitive in nature it began to diverge owing to reproductive isolation.

Despite the strong correlation between DNA relatedness and fertility exhibited in Table 1, exceptions may occur and whole genome DNA complementarity should therefore be regarded as a strong, but not infallible, indicator of biological relatedness. Exceptions to this trend would include chromosomal changes that affect fertility between strains such as inversions, translocations, autopolyploidy, and amphidiploidy (allopolyploidy). Most of these changes would not be detectable in whole genome DNA comparisons. However, amphidiploidy is recognizable if parental species have been included in the study or should be suspected if genome sizes vary between strains showing substantial relatedness. For example, *Saccharomyces carlsbergensis* shows relatively high DNA homology with both *S. cerevisiae* and *S. bayanus*, even though these latter two species demonstrate little relatedness (Figure 1). A comparison of genome sizes suggests *S. carlsbergensis* to be a partial amphidiploid which arose as a natural hybrid of *S. cerevisiae* and *S. bayanus*.[64] In this case, we would predict *S. carlsbergensis* to be infertile with the proposed parents despite relatively high DNA relatedness.

E. Comparisons of Species

Bak and Stenderup[4] appear to have been the first to make extensive use of DNA reassociation as a means for determining relatedness among yeasts. Their study included several phenotypically similar *Candida* species of medical interest. Specifically, their data showed sufficiently high base sequence complementarity (about 65 to 81%) to regard *C. albicans*, *C. stellatoidea*, and *C. claussenii* as conspecific.

Additional advances in the clarification of imperfect taxa through DNA studies were made by Meyer and Phaff,[65] who found high degrees of homology between *C. lusitaniae* and *C.*

Table 2
EXTENT OF DNA RELATEDNESS BETWEEN *SACCHAROMYCES ROSEI* AND PHENOTYPICALLY SIMILAR SPECIES OF *SACCHAROMYCES* NOW CLASSIFIED IN *TORULASPORA*

Species	DNA reassociation with *S. rosei*[a] (%)
S. saitoanus	100.2
S. inconspicuus	99.9
Debaryomyces nilssonii	99.9
S. florenzani	99.6
S. fermentati	99.5
S. microellipsoides var. *osmophilus*	99.1
S. vafer	98.4
S. delbrueckii	98.0
S. pretoriensis	13.3
S. microellipsoides var. *microellipsoides*	10.7
S. kloeckerianus	10.5
S. cerevisiae	7.7
S. eupagycus	5.8
S. montanus	5.7
S. florentinus	5.5
S. bisporus	3.5

[a] Data from Price, C. W., Fuson, G. B., and Phaff, H. J., *Microbiol. Rev.*, 42, 161, 1978.

obtusa and between *C. salmonicola* and *C. sake;* by Meyer et al.,[66] who demonstrated *C. cloacae* and *C. subtropicalis* to be synonyms of *C. maltosa;* and by Kurtzman et al.,[30] who showed *C. krusei* and *C sorbosa* to represent, respectively, the imperfect states of *Issatchenkia orientalis* and *I. occidentalis.* Kurtzman et al.[67] and Manachini[68] demonstrated *C. utilis* to be the imperfect form of *Pichia (Hansenula) jadinii.* This pairing is of particular interest, because some strains of *C. utilis* are used as food yeasts while others are from animal sources, and the teleomorph *P. jadinii* was isolated from a human infection. In view of this, strain histories become quite important when selecting cultures for certain uses. The work of Mendonça-Hagler and Phaff[34] provided clear evidence that *Saccharomyces telluris, Torulopsis bovina, T. pintolopesii,* and *C. sloofii* are conspecific. This study also convincingly demonstrated that pseudomycelium formation provides no distinction between *Torulopsis* and *Candida.*

Further clarification of speciation in ascosporogenous yeasts resulted from the work of Price et al.[5] Of particular interest were the rough-spored species of *Saccharomyces* which are now classified in *Torulaspora.*[69] Nine of these species, which included the well-known *S. rosei* and *S. delbrueckii,* showed 98% or greater DNA relatedness and must be considered conspecific (Table 2). All now are classified as *T. delbrueckii.*

In an ultrastructural comparison of ascospores from species assigned to *Pichia* and *Debaryomyces,* Kurtzman and Smiley[70] observed spores of *P. polymorpha* and *P. pseudopolymorpha* to be typical of those found in *Debaryomyces,* i.e., spheroidal and roughened. They suggested *P. polymorpha* and *D. cantarellii* to be conspecific because of this and other phenotypic similarities. This premise was confirmed by Price et al.[5] (Table 3), and their DNA studies further showed *D. phaffii* also a member of the taxon now designated *D. polymorphus.*[71]

Martini and Kurtzman[64] examined the DNA relatedness of 24 species assigned to the

Table 3
EXTENT OF DNA RELATEDNESS BETWEEN *DEBARYOMYCES POLYMORPHUS* AND PHENOTYPICALLY SIMILAR SPECIES WITH ROUGHENED ASCOSPORES

Species	DNA reassociation with *D. polymorphus*[a] (%)
D. cantarellii	96.4
D. phaffii	96.1
D. vanriji	23.9
D. formicarius	23.7
Pichia pseudopolymorpha	21.0
D. castellii	19.8
D. hansenii	8.0
P. terricola	1.9

[a] Data from Price, C. W., Fuson, G. B., and Phaff, H. J., *Microbiol. Rev.*, 42, 161, 1978.

genus *Saccharomyces* sensu stricto. Initial species separations had been made primarily on the basis of fermentation and assimilation of various carbon compounds. Attempts to clarify species on the basis of cell wall antigens,[72] proton magnetic resonance (PMR) spectra of extracted mannans,[73] requirements for vitamins[74] and amino compounds,[75] and through numerical analysis[76] have been largely unsuccessful. The DNA comparisons showed these 24 taxa to be comprised of four distinct species: *S. cerevisiae*, *S. bayanus*, *S. carlsbergensis*, and *S. kluyveri* (Tables 4 and 5). As discussed earlier, *S. carlsbergensis* showed intermediate relatedness between *S. cerevisiae* and *S. bayanus* and appears to be a partial amphidiploid which may have arisen from a natural hybridization between *S. cerevisiae* and *S. bayanus*. Furthermore, the *S. bayanus* parent of this hybrid was probably a strain formerly classified as *S. uvarum*, because both *S. carlsbergensis* and *S. uvarum* ferment melibiose, whereas strains originally classified as *S. bayanus* did not ferment this sugar.

III. MITOCHONDRIAL DNA RELATEDNESS

Relatedness among species of two yeast genera has been examined through comparisons of fragment patterns of mitochondrial DNA (mtDNA) generated with restriction endonucleases. The small size of the mitochondrial genome makes such comparisons practical, whereas the larger nuclear genome would present more fragments than could be reasonably handled. McArthur and Clark-Walker[77] used mtDNA restriction patterns to correlate perfect-imperfect relationships between species in *Dekkera* and *Brettanomyces*. They found identical restriction patterns for the pair *D. bruxellensis*/*B. lambicus*, as well as for the imperfects *B. abstinens*/*B. custersii* and *B. anomalus*/*B. clausenii*. This strongly suggests the pairs to be conspecific. Size differences in mtDNA among the other species assigned to these genera prevented an unambiguous assessment of their relationships.

One potential difficulty with restriction pattern analysis is that mtDNA polymorphisms, which arise from insertions or deletions, will give the erroneous appearance of greater sequence divergence than really exists. Because the rate of change in mtDNA in some organisms may be up to tenfold more rapid than that of nuclear DNA, the resolution afforded by mtDNA patterns may not be sufficient to recognize the more divergent strains of a species.[78,79]

Table 4
EXTENT OF DNA REASSOCIATION BETWEEN *SACCHAROMYCES CEREVISIAE* NRRL Y-12632^{T} [a] AND OTHER SPECIES OF *SACCHAROMYCES* SENSU STRICTO

Species	NRRL no.	DNA reassociation[b] (%)
S. aceti	Y-12617^{T}	96
S. beticus	Y-12625^{T}	99
S. capensis	YB-4237^{T}	94
	Y-12629	99
S. cerevisiae	Y-2034	91
S. chevalieri	Y-12633^{T}	96
S. cordubensis	Y-12636^{T}	100
S. coreanus	Y-12637^{T}	94
S. diastaticus	Y-2416	92
S. gaditensis	Y-12644^{T}	95
S. hienipiensis	Y-6677^{T}	95
S. hispalensis	Y-11846^{T}	94
S. italicus	Y-12649^{T}	96
S. norbensis	Y-12656^{T}	90
S. oleaceus	Y-12657^{T}	94
S. oleaginosus	Y-6679^{T}	89
	Y-12659	86
S. prostoserdovii	Y-6680^{T}	93
	Y-12660	101
S. carlsbergensis	Y-12693^{T}	57
S. abuliensis	Y-11845^{T}	15
S. bayanus	Y-12624^{T}	5
S. globosus	Y-12645^{T}	3
	Y-12646	8
S. heterogenicus	Y-1354^{T}	20
S. inusitatus	Y-12648^{T}	24
S. kluyveri	Y-12651^{T}	−3
S. uvarum	Y-12663^{T}	14
Debaryomyces melissophilus	Y-7585^{T}	6

[a] T = type strain.
[b] Data from Martini, A. V. and Kurtzman, C. P., *Int. J. Syst. Bacteriol.*, 35, 508, 1985.

IV. RIBOSOMAL RNA RELATEDNESS

The rather narrow resolution afforded by whole genome DNA reassociation and mtDNA restriction analysis does not allow verification of species assignments within genera or an understanding of intergeneric relationships. The DNA coding for ribosomal RNA (rRNA) appears to be among the most highly conserved sequences known, and it offers a means for assessing affinities above the species level. One of the most common methods for measuring rRNA complementarity consists of immobilizing single-stranded DNA onto nitrocellulose filters and incubating the filters in a buffer solution with radiolabeled rRNA.[13,24,80] A modification of this method was developed by Baharaeen et al.[81] in which complementary DNA was synthesized on 25S rRNA fragments and then allowed to hybridize with rRNA in solution.

Woese and collaborators (Fox et al.[82]) have effectively assessed phylogenetic relationships among the prokaryotes through cataloging 16S rRNA oligonucleotides generated with T_1

Table 5
EXTENT OF DNA REASSOCIATION BETWEEN *SACCHAROMYCES BAYANUS* NRRL Y-12624^{T} [a] AND *SACCHAROMYCES* SENSU STRICTO SPECIES SHOWING LOW DNA RELATEDNESS WITH *S. CEREVISIAE*

Species	NRRL no.	DNA reassociation[b] (%)
S. abuliensis	Y-11845^{T}	91
S. globosus	Y-12645^{T}	104
	Y-12646	87
S. heterogenicus	Y-1354^{T}	86
S. inusitatus	Y-12648^{T}	94
S. uvarum	Y-12663^{T}	98
S. carlsbergensis	Y-12693^{T}	72
S. kluyveri	Y-12651^{T}	24
Debaryomyces melissophilus	Y-7585^{T}	1

[a] T = type strain.
[b] Data from Martini, A. V. and Kurtzman, C. P., *Int. J. Syst. Bacteriol.*, 35, 508, 1985.

Table 6
EXTENT OF NUCLEAR DNA AND RIBOSOMAL RNA RELATEDNESS AMONG SELECTED YEASTS[a]

Species	Relatedness with *Saccharomyces cerevisiae* (%)	
	DNA	rRNA
Saccharomyces cerevisiae	100	100
S. bayanus (S. uvarum)	40	100
Torulaspora globosa (S. kloeckerianus)	20	97
Zygosaccharomyces rouxii	0	94
Kluyveromyces wickerhamii	0	87
Pichia (Hansenula) jadinii (Candida utilis)	0	77
P. pastoris	0	50

[a] Data from Bicknell, J. N. and Douglas, H. C., *J. Bacteriol.*, 101, 505, 1970.

ribonuclease. Johnson and Harich[83] reported good correlation of values obtained by this method with those from experiments in which rRNAs are hybridized to DNA bound on nitrocellulose. Application of the oligonucleotide cataloging method to yeasts is unlikely in the near future because of the greater number of fragments generated from the 18S ribosomal subunit of eukaryotes.[84]

Bicknell and Douglas[13] were among the first to employ rRNA comparisons to assess phylogenetic relatedness among yeasts. Their study focused on the genus *Saccharomyces*, some species of which now have been assigned to the genera *Zygosaccharomyces*, *Kluyveromyces*, and *Torulaspora*. In general, the more highly related species clustered into groups corresponding to present generic assignments (Table 6). Similar results were obtained for this same group of species by Adoutte-Panvier et al.[85] through electrophoretic and immunochemical comparisons of ribosomal proteins. Bicknell and Douglas also showed that 25S rRNA hybridizations cannot resolve the phylogeny of closely related species because of the highly conserved nature of the sequences. In studies of other yeast species, Segal and Eylan[86-88] reached similar conclusions.

The preceding studies have estimated relatedness from the extent of complementarity of 25S rRNA. Another approach has been to compare the actual sequence of 5S rRNA nucleotides. Hori[89] sequenced the 5S rRNA from a number of prokaryotes and eukaryotes, including several yeasts. Using nucleotide substitution as an evolutionary clock, Hori suggested that prokaryotes and eukaryotes diverged approximately 2.5×10^9 years ago. Walker and Doolittle[90,91] also used 5S rRNA sequences to investigate relatedness among the basidiomycetes. Their work identified two distinct clusters that correlated with the presence or absence of septal dolipores rather than with the traditional separation of these species into the classes Heterobasidiomycetae and Homobasidiomycetae. Further, the data suggest that capped dolipores evolved from capless dolipores which may have evolved from single septal pores. Walker's[92] comparisons of 5S rRNA sequences among the lower ascomycetes suggest this group also to be highly divergent. The sequencing of selected portions of 16S rRNA has revealed evolutionary affinities among the bacteria,[93] and this technique might be successfully applied to yeasts. Primer sequences are used to select for particular stretches of the rRNA molecule and the rates of base substitution are then compared. The lengths of 16S rRNA retrieved are generally several times longer than the 5S rRNA molecule and, therefore, potentially offer more information. If the rRNA sequences selected contain both highly conserved as well as less conserved areas, it may be possible to deduce inter- and intrageneric relationships.

V. PROTEINS AS INDICATORS OF RELATEDNESS

As we have seen, whole genome DNA complementarity is too specific for detecting intrageneric relationships, whereas rRNA comparisons may be too broad in resolution to detect closely related species.[13] However, the DNA coding for certain enzymes such as glutamine synthetase and superoxide dismutase, although less conserved than that of coding for rRNA, is still less divergent than is apparent from comparisons of whole genome DNA.[94,95] By determining changes in the amino acid sequence of these enzymes through such techniques as quantitative microcomplement fixation, an estimate of intermediate relatedness may be found. Immunological studies of protein similarity are very sparse for fungi, and the resultant information is highly dependent on the kinds of proteins studied. Lachance and Phaff[96] compared exo-β-glucanase from species of *Kluyveromyces*, but found that the enzyme was poorly conserved and immunological distances between most species were too great to be reliable for determining evolutionary relationships in that genus.

Electrophoresis of allozymes presents another means for estimating molecular diversity and, as with immunological studies, resolution depends on the extent of sequence conservation. The proportion of point mutations that are electrophoretically detectable is estimated at approximately 0.27, owing to the redundancy in the genetic code and the large proportion of amino acids that are electrically neutral.[97,99] The nearly universal occurrence of extensive protein polymorphism in natural populations has led some workers to believe that variation would prove physiologically important and, therefore, be under selective control; others regard it as without phenotypic effect and thus selectively neutral.[98] Regardless, data derived by this technique do provide insight into evolutionary processes and taxonomy. Baptist and Kurtzman[100] utilized comparative enzyme patterns to separate sexually active strains of *Cryptococcus laurentii* var. *laurentii* from nonreactive strains and from the varieties *magnus* and *flavescens*. Holzschu[99] recently studied the evolutionary relationships among some 400 strains of various cactophilic species of *Pichia*. His study of the banding patterns in starch gels of 14 metabolic enzymes allowed an estimation of the genetic distances among the various yeast populations.

Table 7
DNA RELATEDNESS BETWEEN YEAST SPECIES DIFFERING IN TRADITIONAL TAXONOMIC CHARACTERISTICS

Species	Characteristic/ + or −	DNA relatedness (%)	Ref.
Candida slooffii	Pseudohyphae/ +	80	34
Torulopsis pintolopesii	/ −		
Hansenula wingei	True hyphae/ +	78	110
H. canadensis	/ −		
Debaryomyces formicarius	Glucose Ferm/ +	96	5
D. vanriji	/ −		
Schwanniomyces castellii	Lactose Assim/ +	97	5
S. occidentalis	/ −		
H. minuta	Nitrate Assim/ +	75	107
Pichia lindneri	/ −		
Sterigmatomyces halophilus	Nitrate Assim/ +	100	108
S. indicus	/ −		

VI. COMPARISON OF RELATEDNESS FROM NUCLEIC ACID STUDIES WITH THAT DETERMINED BY OTHER METHODOLOGIES

In recent years it has become apparent that many of the criteria traditionally used for speciation of yeasts have their basis in only a very small portion of the genome. As is clear from genetic studies,[101-104] the assimilation and fermentation of carbon compounds are frequently under the control of single genes. Consequently, carbon utilization tests should be used as a convenient laboratory means for separating taxa rather than as the foundation upon which they are defined. This is also true for the nitrate assimilation reaction. Ability to assimilate nitrate has been of cardinal importance in the definition of many species and genera, since Stelling-Dekker[105] first proposed its use over 50 years ago for the separation of *Hansenula* (NO_3 assim.$^+$) from *Pichia* (NO_3 assim.$^-$). Genetic studies show nitrate assimilation to be under the control of only a few genes.[106] Recently, Kurtzman[107] found 75% DNA relatedness between *P. lindneri* and *H. minuta* and proposed that the two genera be combined. Kurtzman et al.[108] also determined that the nitrate-positive *Sterigmatomyces halophilus* shared 100% DNA base sequence complementarity with the nitrate-negative *S. indicus* (Table 7).

Nucleic acids studies have not proved entirely supportive of conclusions drawn from numerical analysis, from proton magnetic resonance, or from serology of cell-wall mannans. Using numerical analysis, Campbell[109] proposed reduction of *H. wingei, H. canadensis,* and both varieties of *H. bimundalis* to a single species. Fuson and co-workers[110] showed from DNA comparisons that *H. canadensis* and *H. wingei* are conspecific, while this latter species and *H. bimundalis* are unrelated. Kurtzman[111] demonstrated only 21% complementarity between the two varieties of *H. bimundalis*. This disparity of the two methods probably results from the relatively limited amount of phenotypic data available for numerical analysis, while DNA studies have the whole genome as a data base. This also appears to be the case for separations made by means of proton magnetic resonance; Spencer and Gorin[112] found the spectra of cell-wall mannans from *H. beckii* and *H. canadensis* to be quite similar, but different from the pattern shared by *H. wingei* and the varieties of *H. bimundalis*. In further comparisons of this group, Tsuchiya et al.[113] reported the cell surface antigens of *H. wingei, H. canadensis,* and *H. beckii* to be indistinguishable. The reason for this discrepancy is not at all clear, but Ballou[114] has shown for *Saccharomyces* mannans that single gene changes can impart differences in immunological reaction. The difference between the two main PMR spectral types of mannan in *S. cerevisiae* was also found to be controlled by a single gene.[115]

Reliance upon the presence or absence of pseudohyphae and true hyphae for taxonomic assignment has been questioned. Van Uden and Buckley[116] as well as Yarrow and Meyer[117] have pointed out that some taxa do not consistently form pseudohyphae. For primarily this reason, these latter authors transferred species of *Torulopsis* to *Candida*. Fuson et al.[110] provided evidence that the presence or absence of hyphal development is not necessarily a good trait for separation of species. This conclusion was based on the observation that the type strain of *H. canadensis* which is differentiated from *H. wingei* by its homothallism, ascospore number, and lack of true hyphae has a higher degree of DNA homology (77.6%) with the type strain of *H. wingei* than do some strains identified as *H. wingei* (62.2%).

Differentiation of a number of genera is dependent upon ascospore shape and surface structure. An opportunity to test the soundness of this convention in the genera *Schwanniomyces, Saccharomyces, Debaryomyces,* and *Pichia* arose through a comparison of the DNA data of Price et al.[5] with the scanning electron microscopy studies by Kurtzman and colleagues.[70,118-121] All four species of *Schwanniomyces* had high DNA relatedness and similar spore architecture. The least related, *S. persoonii* at 80% complementarity, showed fewer and less pronounced spore protuberances than the other three species,[120,122] all four of which are now regarded as conspecific. Ascospore topography for many species assigned to *Torulaspora* was essentially identical and also frequently indistinguishable from the majority of species in *Debaryomyces*. Although many species within their respective genera were found conspecific by DNA homology, others proved distinct despite ascospore similarity. The few species with recognizably different spores showed little DNA relatedness. Thus, species with unlike spores can be expected to show little DNA complementarity, but no prediction can be made concerning species with similar spores. A few exceptions exist among other groups. Ascospores of *P. ohmeri* may be either spheroidal or hat-shaped, depending upon mating type,[123,124] and spore shape in *Yarrowia lipolytica* is mating type-dependent.[125]

Kreger-van Rij and Veenhuis[126] demonstrated by transmission electron microscopy a basic difference in hyphal septa among ascomycetous yeasts. Three classes of septum ultrastructure were observed: a single central pore or closure line, plasmodesmata, and a dolipore configuration. All mycelial species of *Pichia* (including *Hansenula*) have the central pore, while plasmodesmata are common to all *Saccharomycopsis* species except *(Y.) S. lipolytica*. Dolipores were found in species of *Ambrosiozyma*. Comparison of these species by molecular techniques which detect distant relatedness might show ultrastructural differences in septa to be of evolutionary significance.

The analysis of coenzyme Q or ubiquinone by Yamada and co-workers[127-136] has revealed the number of isoprene subunits to vary from five to ten (Q-5 to Q-10) among yeast taxa. In many genera, all species have the same Q number, but heterogeneity exists in such genera as *Pichia*. This may indicate that the atypical species are incorrectly assigned, and a comparison of these species by molecular techniques could have interesting phylogenetic implications.

VII. CONCLUDING REMARKS

As we have seen from numerous examples, application of molecular techniques has provided insight into evolutionary relationships among the yeasts that were, in many cases, previously unsuspected. These new findings have changed many of our older concepts, and this has resulted in a considerable number of taxonomic revisions. While these name changes are inconvenient and may even seem capricious, they are necessary if our taxonomy is to reflect phylogeny, a supposition implicit in the assignment of species to genera and of genera to higher orders of classification.

REFERENCES

1. **Ferenczy, L.,** Microbial protoplast fusion, in *Genetics as a Tool in Microbiology*, Glover, S. W. and Hopwood, D. A., Eds., Cambridge University Press, Cambridge, England, 1918, 1.
2. **Gunge, N.,** Yeast DNA plasmids, *Ann. Rev. Microbiol.*, 37, 253, 1983.
3. **Toh-e, A., Araki, H., Utatsu, I., and Oshima, Y.,** Plasmids resembling 2-μm DNA in the osmotolerant yeasts *Saccharomyces bailii* and *Saccharomyces bisporus*, *J. Gen. Microbiol.*, 130, 2527, 1984.
4. **Bak, A. L. and Stenderup, A.,** Deoxyribonucleic acid homology in yeasts. Genetic relatedness within the genus *Candida*, *J. Gen. Microbiol.*, 59, 21, 1969.
5. **Price, C. W., Fuson, G. B., and Phaff, H. J.,** Genome comparison in yeast systematics: delimitation of species within the genera *Schwanniomyces, Saccharomyces, Debaryomyces,* and *Pichia*, *Microbiol. Rev.*, 42, 161, 1978.
6. **Kurtzman, C. P., Smiley, M. J., Johnson, C. J., Wickerham, L. J., and Fuson, G. B.,** Two new and closely related heterothallic species. *Pichia amylophilia* and *Pichia mississippiensis:* characterization by hybridization and deoxyribonucleic acid reassociation, *Int. J. Syst. Bacteriol.*, 30, 208, 1980.
7. **Lipinski, C., Ferro, A. J., and Mills, D.,** Macromolecule synthesis in a mutant of *Saccharomyces cerevisiae* inhibited by *S*-adenosylmethionine, *Mol. Gen. Genet.*, 144, 301, 1976.
8. **Van Etten, J. L. and Freer, S. N.,** Simple procedure for disruption of fungal spores, *Appl. Environ. Microbiol.*, 35, 622, 1978.
9. **Martini, A. and Phaff, H. J.,** The optical determination of DNA-DNA homologies in yeasts, *Ann. Microbiol.*, 23, 59, 1973.
10. **Storck, R. and Alexopoulos, C. J.,** Deoxyribonucleic acid of fungi, *Bacteriol. Rev.*, 34, 126, 1970.
11. **Mendonça-Hagler, L. C., Travassos, L. R., Lloyd, K. O., and Phaff, H. J.,** Deoxyribonucleic acid base composition and hybridization studies on the human pathogen *Sporothrix schenckii* and *Ceratocystis* species, *Infect. Immun.*, 8, 674, 1974.
12. **Meyer, S. A. and Phaff, H. J.,** Deoxyribonucleic acid base composition in yeasts, *J. Bacteriol.*, 97, 52, 1969.
13. **Bicknell, J. N. and Douglas, H. C.,** Nucleic acid homologies among species of *Saccharomyces*, *J. Bacteriol.*, 101, 505, 1970.
14. **Smith, D. and Halvorson, H. O.,** The isolation of DNA from yeast, in *Methods in Enzymology*, Vol. 12, Part A, Grossman, L. and Moldave, K., Eds., Academic Press, New York, 1967, 538.
15. **Cryer, D. R., Eccleshall, R., and Marmur, J.,** Isolation of yeast DNA, in *Methods in Cell Biology*, Vol. 12, Prescott, D. M., Ed., Academic Press, New York, 1975, 39.
16. **Brady, R. J.,** A technique for preparation of spheroplasts of *Rhodotorula*, *Am. Soc. Microbiol.*, K65, 148, 1981.
17. **Rhodes, J. C. and Kwon-Chung, K. J.,** A new efficient method for protoplast formation in *Cryptococcus neoformans*, *Am. Soc. Microbiol.*, F61, 336, 1982.
18. **Marmur, J.,** A procedure for the isolation of DNA from microorganisms, *J. Mol. Biol.*, 3, 208, 1961.
19. **Bernardi, G., Faures, M., Piperno, G., and Slonimski, P. P.,** Mitochondrial DNAs from respiratory-sufficient and cytoplasmic respiratory-deficient mutants of yeast, *J. Mol. Biol.*, 48, 23, 1970.
20. **Holzschu, D. L. and Phaff, H. J.,** personal communication, 1980.
21. **Holzschu, D. L., Tredick, J., and Phaff, H. J.,** Validation of the yeast *Sporidiobolus ruinenii* based on its deoxyribonucleic acid relatedness to other species of the genus *Sporidiobolus*, *Curr. Microbiol.*, 5, 73, 1981.
22. **Kurtzman, C. P., Smiley, M. J., and Axt, A.,** unpublished data, 1981.
23. **Timberlake, W. E.,** Low repetitive DNA content in *Aspergillus nidulans*, *Science*, 202, 973, 1978.
24. **Johnson, J. L.,** Genetic characterization, in *Manual of Methods for General Bacteriology*, Gerhardt, P., Ed., American Society for Microbiology, Washington, D.C., 1981, 450.
25. **Gillespie, D. and Spiegelman, S.,** A quantitative assay for DNA-RNA hybrids with DNA immobilized on a membrane, *J. Mol. Biol.*, 12, 829, 1965.
26. **Kirby, K. S.,** Isolation of nucleic acids with phenolic solvents, *Methods Enzymol.*, 12B, 87, 1968.
27. **Marmur, J. and Doty, P.,** Determination of the base composition of DNA from its thermal denaturation temperature, *J. Mol. Biol.*, 5, 109, 1962.
28. **Bak, A. L.,** DNA base composition in mycoplasma, bacteria, and yeast, *Curr. Top. Microbiol. Immunol.*, 61, 89, 1973.

29. **Schildkraut, C. L., Marmur, J., and Doty, P.,** Determination of the base composition of deoxyribonucleic acid from its buoyant density in CsCl, *J. Mol. Biol.*, 4, 430, 1962.
30. **Kurtzman, C. P., Smiley, M. J., and Johnson, C. J.,** Emendation of the genus *Issatchenkia* Kurdriavzev and comparison of species by deoxyribonucleic acid reassociation, mating reaction, and ascospore ultrastructure, *Int. J. Syst. Bacteriol.*, 30, 503, 1980.
31. **Meyer, S. A., Smith, M. T., and Simione, F. P., Jr.,** Systematics of *Hanseniaspora* Zikes and *Kloeckera* Janke, *Antonie van Leeuwenhoek: J. Microbiol. Serol.*, 44, 79, 1978.
32. **Meyer, S. A. and Phaff, H. J.,** Taxonomic significance of the DNA base composition in yeasts, in *Recent Trends in Yeast Research*, Vol. 1, Ahearn, D. G., Ed., Spectrum, Georgia State University, Atlanta, 1970, 1.
33. **Commerford, S. L.,** Iodination of nucleic acids *in vitro*, *Biochemistry*, 10, 1993, 1971.
34. **Mendonça-Hagler, L. C. and Phaff, H. J.,** Deoxyribonucleic acid base composition and DNA/DNA hybrid formation in psychrophobic and related yeasts, *Int. J. Syst. Bacteriol.*, 25, 222, 1975.
35. **Holzschu, D. L., Presley, H. L., Miranda, M., and Phaff, H. J.,** Identification of *Candida lusitaniae* as an opportunistic yeast in humans, *J. Clin. Microbiol.*, 10, 202, 1979.
36. **Kelly, R. B., Cozzrelli, N. R., Deutscher, M. P., Lehman, I. R., and Kornberg, A.,** Enzymatic synthesis of deoxyribonucleic acids, *J. Biol. Chem.*, 245, 39, 1970.
37. **Maniatis, T., Jeffrey, A., and Kleid, D. G.,** Nucleotide sequence of the rightward operator of phage, *Proc. Natl. Acad. Sci. U.S.A.*, 72, 1184, 1975.
38. **Denhardt, D. T.,** A membrane filter technique for the detection of complementary DNA, *Biochem. Biophys. Res. Commun.*, 23, 641, 1966.
39. **Brenner, D. J.,** Deoxyribonucleic acid reassociation in the taxonomy of enteric bacteria, *Int. J. Syst. Bacteriol.*, 23, 298, 1973.
40. **Crosa, J. H., Brenner, D. J., and Falkow, S.,** Use of a single-strand-specific nuclease for analysis of bacterial and plasmid deoxyribonucleic acid homo- and heteroduplexes, *J. Bacteriol.*, 115, 904, 1973.
41. **Subirana, J. A. and Doty, P.,** Kinetics of renaturation of renatured DNA. I. Spectrophotometric results, *Biopolymers*, 4, 171, 1966.
42. **De Ley, J., Cattoir, H., and Reynaerts, A.,** The quantitative measurement of DNA hybridization from renaturation rates, *Eur. J. Biochem.*, 12, 133, 1970.
43. **Seidler, R. J. and Mandel, M.,** Quantitative aspects of DNA renaturation: DNA base composition, state of chromosome replication, and polynucleotide homologies, *J. Bacteriol.*, 106, 608, 1971.
44. **Ullman, J. S. and McCarthy, B. J.,** The relationships between mismatched base pairs and the thermal stability of DNA duplexes. I. Effects of depurination and chain scission, *Biochim. Biophys. Acta*, 294, 405, 1973.
45. **Brenner, D. J., Steigerwalt, A. G., and Fanning, G. R.,** Differentiation of *Enterobacter aerogenes* from klebsiellae by deoxyribonucleic acid reassociation, *Int. J. Syst. Bacteriol.*, 22, 193, 1972.
46. **McCarthy, B. J. and Church, R. B.,** The specificity of molecular hybridization reactions, *Annu. Rev. Biochem.*, 39, 131, 1970.
47. **Christiansen, C., Bak, A. L., Stenderup, A., and Christiansen, G.,** Repetitive DNA in yeast, *Nature (London) New Biol.*, 231, 176, 1971.
48. **Britten, R. J. and Kohne, D. E.,** Repeated sequences in DNA, *Science*, 161, 529, 1968.
49. **Bak, A. L., Christiansen, C., and Christiansen, G.,** Circular, repetitive DNA in yeast, *Biochim. Biophys. Acta*, 269, 527, 1972.
50. **O'Connor, R. M., McArthur, C. R., and Clark-Walker, G. D.,** Closed-circular DNA from mitochondrial-enriched fractions of four *petite*-negative yeasts, *Eur. J. Biochem.*, 53, 137, 1975.
51. **O'Connor, R. M., McArthur, C. R., and Clark-Walker, G. D.,** Respiratory-deficient mutants of *Torulopsis glabrata*, a yeast with circular mitochondrial deoxyribonucleic acid of 6 μm, *J. Bacteriol.*, 126, 959, 1976.
52. **Gunge, N., Tamuru, A., Ozawa, F., and Sakaguchi, K.,** Isolation and characterization of linear deoxyribonucleic acid plasmids from *Kluyveromyces lactis* and the plasmid-associated killer character, *J. Bacteriol.*, 145, 382, 1981.
53. **Petes, T. D., Beyers, B., and Fangman, W. L.,** Size and structure of yeast chromosomal DNA, *Proc. Natl. Acad. Sci. U.S.A.*, 70, 3072, 1973.
54. **Hartwell, L. H.,** *Saccharomyces cerevisiae* cell cycle, *Bacteriol. Rev.*, 38, 164, 1974.
55. **Blamire, J., Cryer, D. R., Finkelstein, D. B., and Marmur, J.,** Sedimentation properties of yeast nuclear and mitochondrial DNA, *J. Mol. Biol.*, 67, 11, 1972.
56. **Hollenberg, C. P., Borst, P., and van Brugger, E. F. J.,** Mitochondrial DNA. V. A 25-μ closed circular duplex molecule in wild-type yeast mitochondria. Structure and kinetic complexity, *Biochim. Biophys. Acta*, 209, 1, 1970.
57. **Williamson, D. H. and Fennell, D. J.,** The use of fluorescent DNA-binding agent for detecting and separating yeast mitochondria DNA, in *Methods in Cell Biology*, Vol. 12, Prescott, D. M., Ed., Academic Press, New York, 1975, 335.

58. **Marmur, J., Rownd, R., and Schildkraut, C. L.,** Denaturation and renaturation of DNA, *Prog. Nucl. Acid Res. Mol. Biol.*, 1, 231, 1963.
59. **Johnson, J. L.,** Use of nucleic acid homologies in the taxonomy of anaerobic bacteria, *Int. J. Syst. Bacteriol.*, 23, 308, 1973.
60. **van der Walt, J. P. and Johannsen, E.,** A comparison of interfertility and in vitro DNA-DNA reassociation as criteria for speciation in the genus *Kluyveromyces*, *Antonie van Leeuwenhoek; J. Microbiol. Serol.*, 45, 281, 1979.
61. **Johannsen, E.,** Hybridization studies within the genus *Kluyveromyces* van der Walt emend. van der Walt, *Antonie van Leeuwenhoek; J. Microbiol. Serol.*, 46, 177, 1980.
62. **Martini, A.,** Ibridazioni DNA/DNA tra specie di lieviti del genere *Kluyveromyces*, *Ann. Fac. Agrar. Univ. Studi. Perugia*, 28, 1, 1973.
63. **Phaff, H. J., Miller, M. W., and Miranda, M.,** *Pichia scutulata*, a new species from tree exudates, *Int. J. Syst. Bacteriol.*, 26, 326, 1976.
64. **Martini, A. V. and Kurtzman, C. P.,** Deoxyribonucleic acid relatedness among species of the genus *Saccharomyces* sensu stricto, *Int. J. Syst. Bacteriol.*, 35, 508, 1985.
65. **Meyer, S. A. and Phaff, H. J.,** DNA base composition and DNA-DNA homology studies as tools in yeast systematics, in *Yeasts, Models in Science and Technics*, Kochová-Kratochvilová, A. and Minarik, E., Eds., Publishing House, Slovak Academy of Science, Bratislava, Czechoslovakia, 1972, 375.
66. **Meyer, S. A., Anderson, K., Brown, R. E., Smith, M. Th., Yarrow, D., Mitchell, G., and Ahearn, D. G.,** Physiological and DNA characterization of *Candida maltosa*, a hydrocarbon utilizing yeast, *Arch. Microbiol.*, 104, 225, 1975.
67. **Kurtzman, C. P., Johnson, C. J., and Smiley, M. J.,** Determination of conspecificity of *Candida utilis* and *Hansenula jadinii* through DNA reassociation, *Mycologia*, 71, 844, 1979.
68. **Manachini, P. L.,** DNA sequence similarity, cell wall mannans, and physiological characteristics in some strains of *Candida utilis*, *Hansenula jadinii*, and *Hansenula petersonii*, *Antonie van Leeuwenhoek; J. Microbiol. Serol.*, 45, 451, 1979.
69. **Yarrow, D.,** *Torulaspora* Lindner, in *The Yeasts, A Taxonomic Study*, 3rd ed., Kreger-van Rij, N. J. W., Ed., Elsevier, Amsterdam, 1984, 434.
70. **Kurtzman, C. P. and Smiley, M. J.,** A taxonomic re-evaluation of the round-spored species of *Pichia*, in *Proc. 4th Int. Symp. on Yeasts*, Part I, Klaushofer, H. and Sleytr, U. B., Eds., Hochschülerschaft an der Hochschule für Bodenkultur, Vienna, Austria, 1974, 231.
71. **Price, C. W. and Phaff, H. J.,** *Debaryomyces polymorphus* and *D. pseudopolymorphus*, new taxonomic combinations, *Mycologia*, 71, 444, 1979.
72. **Fukazawa, J., Shinoda, T., Nishikawa, A., and Nakase, T.,** Synonymy of *Saccharomyces cerevisiae* Hansen 1883 and *Saccharomyces uvarum* Beijerinck 1898: significance of cell wall antigens in yeast classification, *Int. J. Syst. Bacteriol.*, 30, 196, 1980.
73. **Spencer, J. F. T. and Gorin, P. A. J.,** Systematics of the genera *Saccharomyces*, *Schizosaccharomyces*, *Endomycopsis*, *Kluyveromyces*, *Schwanniomyces*, and *Brettanomyces*: proton magnetic resonance spectra of the mannans and mannose-containing polysaccharides as an aid in classification, *Antonie van Leeuwenhoek; J. Microbiol. Serol.*, 35, 361, 1969.
74. **Fiol, J. B.,** Systématique des *Saccharomyces*: osidases et besoins vitaminiques, *Mycopathology*, 58, 19, 1976.
75. **Brady, B.,** Utilization of amino compounds by yeasts of the genus *Saccharomyces*, *Antonie van Leeuwenhoek; J. Microbiol. Serol.*, 31, 95, 1965.
76. **Campbell, J.,** Computer identification of yeasts of the genus *Saccharomyces*, *J. Gen. Microbiol.*, 77, 127, 1973.
77. **McArthur, C. R. and Clark-Walker, G. D.,** Mitochondrial DNA size diversity in the *Dekkera/Brettanomyces* yeasts, *Curr. Genet.*, 7, 29, 1983.
78. **Groot, G. S. P., Flavell, R. A., and Sanders, J. P. M.,** Sequence homology of nuclear and mitochondrial DNAs of different yeasts, *Biochim. Biophys. Acta*, 378, 186, 1975.
79. **Brown, W. M., George, M., Jr., and Wilson, A. C.,** Rapid evolution of animal mitochondrial DNA, *Proc. Natl. Acad. Sci. U.S.A.*, 76, 1967, 1979.
80. **Kennell, D. E.,** Principles and practices of nucleic acid hybridization, *Prog. Nucl. Acid Res. Mol. Biol.*, 11, 259, 1971.
81. **Baharaeen, S., Melcher, U., and Vishniac, H. S.,** Complementary DNA-25S ribosomal RNA hybridization: an improved method for phylogenetic studies, *Can. J. Microbiol.*, 29, 546, 1983.
82. **Fox, G. E., Stackebrandt, E., Hespell, R. B., Gibson, J., Maniloff, J., Dyer, T. A., Wolfe, R. S., Balch, W. E., Tanner, R. S., Magrum, L. J., Zablen, L. B., Balkemore, R., Gupta, R., Bonen, L., Lewis, B. J., Stahl, D. A., Leuhrsen, K. R., Chen, K. N., and Woese, C. R.,** The phylogeny of prokaryotes, *Science*, 209, 457, 1980.
83. **Johnson, J. L. and Harich, B.,** Comparisons of procedures for determining ribosomal ribonucleic acid similarities, *Curr. Microbiol.*, 9, 111, 1983.

84. **Fox, G. E.,** personal communication, 1981.
85. **Adoutte-Panvier, A., Davies, J. E., Gritz, L. R., and Littlewood, B. S.,** Studies of ribosomal proteins of yeast species and their hybrids gel electrophoresis and immunochemical cross-reactions, *Mol. Gen. Genet.*, 179, 273, 1980.
86. **Segal, E. and Eylan, E.,** Genetic relatedness of *Candida albicans* to asporogenous and ascosporogenous yeasts as reflected by nucleic acid homologies, *Microbios*, 9, 25, 1974.
87. **Segal, E. and Eylan, E.,** Nucleic acid homologies between *Candida albicans* and *Hansenula* species, *Microbios*, 10, 133, 1974.
88. **Segal, E. and Eylan, E.,** Nucleic acid homology studies among *Candida albicans, Syringospora albicans*, and *Leucosporidium* species, *Microbios*, 12, 111, 975.
89. **Hori, H.,** Molecular evolution of 5S RNA, *Mol. Gen. Genet.*, 145, 119, 1976.
90. **Walker, W. F. and Doolittle, W. F.,** Redividing the basidiomyces on the basis of 5S rRNA sequences, *Nature*, 229, 723, 1982.
91. **Walker, W. F. and Doolittle, W. F.,** 5S rRNA sequences from eight basidiomycetes and fungi imperfecti, *Nucl. Acids Res.*, 11, 7625, 1983.
92. **Walker, W. F.,** 5S ribosomal RNA sequences from ascomycetes and evolutionary implications, *Syst. Appl. Microbiol.*, 6, 48, 1985.
93. **Woese, C. R., Debrunner-Vossbrinck, B. A., Oyaizu, H., Stackebrandt, E., and Ludwig, W.,** Gram-positive bacteria: possible photosynthetic ancestry, *Science*, 229, 762, 1985.
94. **Baumann, L. and Baumann, P.,** Studies of relationship among terrestrial *Pseudomonas, Alcaligenes*, and enterobacteria by an immunological comparison of glutamine synthetase, *Arch. Microbiol.*, 119, 25, 1978.
95. **Baumann, L., Bang, S. S., and Baumann, P.,** Study of relationship among species of *Vibrio, Photobacterium*, and terrestrial enterobacteria by an immunological comparison of glutamine synthetase and superoxide dismutase, *Curr. Microbiol.*, 4, 133, 1980.
96. **Lachance, M. A. and Phaff, H. J.,** Comparative study of molecular size and structure of exo-β-glucanases from *Kluyveromyces* and other yeast genera: evolutionary and taxonomic implications, *Int. J. Syst. Bacteriol.*, 29, 70, 1979.
97. **Baptist, J. N., Shaw, C. R., and Mandel, M.,** Comparative zone electrophoresis of enzymes of *Pseudomonas solanacearum* and *Pseudomonas cepacia, J. Bacteriol.*, 108, 799, 1971.
98. **Selander, R. K.,** Genetic variation in natural populations, in *Molecular Evolution*, Ayala, F. J., Ed., Sinauer Associates, Sunderland, Mass., 1976, 21.
99. **Holzschu, D. L.,** Molecular Taxonomy and Evolutionary Relationships among Cactophilic Yeasts, Ph.D. thesis, University of California, Davis, 1981.
100. **Baptist, J. N. and Kurtzman, C. P.,** Comparative enzyme patterns in *Cryptococcus laurentii* and its taxonomic varieties, *Mycologia*, 68, 1195, 1976.
101. **Winge, O. and Roberts, C.,** Inheritance of enzymatic characters in yeast, and the phenomenon of long-term adaptation, *Comp. Rend. Trav. Lab. Carlsberg*, 24, 263, 1949.
102. **Lindegren, C. C. and Lindegren, G.,** Unusual gene-controlled combinations of carbohydrate fermentations in yeast hybrids, *Proc. Natl. Acad. Sci. U.S.A.*, 35, 23, 1949.
103. **Kurtzman, C. P. and Smiley, M. J.,** Heterothallism in *Pichia kudriavzevii* and *Pichia terricola, Antonie van Leeuwenhoek; J. Microbiol. Serol.*, 42, 355, 1976.
104. **Starmer, W. T., Phaff, H. J., Miranda, M., and Miller, M. W.,** *Pichia amethionina*, a new heterothallic yeast associated with the decaying stems of cereoid cacti, *Int. J. Syst. Bacteriol.*, 28, 433, 1978.
105. **Stelling-Dekker, N. M.,** *Die sporogenen Hefen, Verh. Kon. Akad. Wetenschap. Amsterdam, Afd. Natuurkunde Sect.* II 28, Amsterdam, Netherlands, 1931, 1.
106. **Marzluf, G. A.,** Regulation of nitrogen metabolism and gene expression in fungi, *Microbiol. Rev.*, 45, 437, 1981.
107. **Kurtzman, C. P.,** Synonomy of the yeast genera *Hansenula* and *Pichia* demonstrated through comparisons of deoxyribonucleic acid relatedness, *Antonie van Leeuwenhoek; J. Microbiol. Serol.*, 50, 209, 1984.
108. **Kurtzman, C. P., Smiley, M. J., Johnson, C. J., and Hoffman, M. J.,** Deoxyribonucleic acid relatedness among species of *Sterigmatomyces, Abstr. 5th Int. Symp. on Yeasts*, 1980, Y-5.2.5(L), 246.
109. **Campbell, I.,** Numerical analysis of *Hansenula, Pichia*, and related yeast genera, *J. Gen. Microbiol.*, 77, 427, 1973.
110. **Fuson, G. B., Price, C. W., and Phaff, H. J.,** Deoxyribonucleic acid sequence relatedness among some members of the yeast genera *Hansenula, Int. J. Syst. Bacteriol.*, 29, 64, 1979.
111. **Kurtzman, C. P.,** Resolution of varietal relationships within the species *Hansenula anomala, Hansenula bimundalis*, and *Pichia nakazawae* through comparisons of DNA relatedness, *Mycotaxon*, 19, 271, 1984.
112. **Spencer, J. F. T. and Gorin, P. A. J.,** Systematics of the genera *Hansenula* and *Pichia:* proton magnetic resonance spectra of their mannans as an aid in classification, *Can. J. Microbiol.*, 15, 375, 1969.
113. **Tsuchiya, T., Fukazawa, Y., Taguchi, M., Nakase, T., and Shinoda, T.,** Serological aspects on yeast classification, *Mycopathol. Mycol. Appl.*, 53, 77, 1974.

114. **Ballou, C. E.,** Some aspects of the structure, immunochemistry, and genetic control of yeast mannans, *Adv. Enzymol.*, 40, 239, 1974.
115. **Spencer, J. F. T., Gorin, P. A. J., and Rank, G. H.,** The genetic control of the two types of mannan produced by *Saccharomyces cerevisiae*, *Can. J. Microbiol.*, 17, 1451, 1971.
116. **van Uden, N. and Buckley, H.,** *Candida* Berkhout, in *The Yeasts, A Taxonomic Study*, 2nd ed., Lodder, J., Ed., North-Holland, Amsterdam, 1970, 893.
117. **Yarrow, D. and Meyer, S. A.,** Proposal for amendment of the diagnosis of the genus *Candida* Berkhout nom. cons, *Int. J. Syst. Bacteriol.*, 28, 611, 1978.
118. **Kurtzman, C. P. and Kreger-van Rij, N. J. W.,** Ultrastructure of ascospores from *Debaryomyces melissophilus*, a new taxonomic combination, *Mycologia*, 68, 422, 1976.
119. **Kurtzman, C. P. and Smiley, M. J.,** Taxonomy of *Pichia carsonii* and its synonyms *Pichia vini* and *P. vini* var. *melibiosi:* comparison by DNA reassociation, *Mycologia*, 71, 658, 1979.
120. **Kurtzman, C. P., Smiley, M. J., and Baker, F. L.,** Scanning electron microscopy of ascospores of *Schwanniomyces*, *J. Bacteriol.*, 112, 1380, 1972.
121. **Kurtzman, C. P., Smiley, M. J., and Baker, F. L.,** Scanning electron microscopy of ascospores of *Debaryomyces* and *Saccharomyces*, *Mycopathol. Mycol. Appl.*, 55, 29, 1975.
122. **Kreger-van Rij, N. J. W.,** Electron microscopy of sporulation in *Schwanniomyces alluvius*, *Antonie van Leeuwenhoek; J. Microbiol. Serol.*, 43, 55, 1977.
123. **Wickerham, L. J. and Burton, K. A.,** A clarification of the relationship of *Candida guilliermondii* to other yeasts by a study of their mating types, *J. Bacteriol.*, 68, 594, 1954.
124. **Fuson, G. B., Price, C. W., and Phaff, H. J.,** Deoxyribonucleic acid base sequence relatedness among strains of *Pichia ohmeri* that produce dimorphic ascospores, *Int. J. Syst. Bacteriol.*, 30, 217, 1980.
125. **Wickerham, L. J., Kurtzman, C. P., and Herman, A. I.,** Sexuality in *Candida lipolytica*, in *Recent Trends in Yeast Research*, Vol. 1, Ahearn, D. G., Ed., Spectrum, Georgia State University, Atlanta, 1970, 81.
126. **Kreger-van Rij, N. J. W. and Veenhuis, M.,** Electron microscopy of septa in ascomycetous yeasts, *Antonie van Leeuwenhoek; J. Microbiol. Serol.*, 39, 481, 1973.
127. **Yamada, Y. and Kondo, K.,** Taxonomic significance of the coenzyme Q system in yeasts and yeast-like fungi (2), in *Fermentation Technology Today, Proc. 4th Int. Fermentation Symp.*, Terui, G., Ed., Society of Fermentation Technology, Osaka, Japan, 1972, 781.
128. **Yamada, Y. and Kondo, K.,** Coenzyme Q system in the classification of the yeast genera *Rhodotorula* and *Cryptococcus*, and the yeast-like genera *Sporobolomyces* and *Rhodosporidium*, *J. Gen. Appl.*, 19, 59, 1973.
129. **Yamada, Y., Arimoto, M., and Kondo, K.,** Coenzyme Q system in the classification of the ascosporogenous yeast genus *Schizosaccharomyces* and yeast-like genus *Endomyces*, *J. Gen. Appl. Microbiol.*, 19, 353, 1973.
130. **Yamada, Y., Kanematsu, Y., Ohashi, M., and Kondo, K.,** On the partly reduced coenzyme Q isolated from "*Rhodotorula lactosa*" IFO 1058 and its relation to the taxonomic position, *Agric. Biol. Chem.*, 37, 621, 1973.
131. **Yamada, Y., Okada, T., Ueshima, O., and Kondo, K.,** Coenzyme Q system in the classification of the ascosporogenous yeast genera *Hansenula* and *Pichia*, *J. Gen. Appl. Microbiol.*, 19, 189, 1973.
132. **Yamada, Y., Arimoto, M., and Kondo, K.,** Coenzyme Q system in the classification of apiculate yeasts in the genera *Nadsonia, Saccharomyces, Hanseniaspora, Kloeckera,* and *Wickerhamia*, *J. Gen. Appl. Microbiol.*, 22, 293, 1976.
133. **Yamada, Y., Nojiri, M., Matsuyama, M., and Kondo, K.,** Coenzyme Q system in the classification of the ascosporogenous yeast genera *Debaryomyces, Saccharomyces, Kluyveromyces,* and *Endomycopsis*, *J. Gen. Appl. Microbiol.*, 22, 325, 1976.
134. **Yamada, Y., Arimoto, M., and Kondo, K.,** Coenzyme Q system in the classification of some ascosporogenous yeast genera in the families Saccharomycetaceae and Spermophthoraceae, *Antonie van Leeuwenhoek; J. Microbiol. Serol.*, 43, 65, 1977.
135. **Yamada, Y., Takinami-Nakamura, H., Tahara, Y., and Smith, M. Th.,** The coenzyme Q system in the classification of the ascosporogenous yeast genus *Dekkera* and the asporogenous yeast genus *Brettanomyces*, *Antonie van Leeuwenhoek; J. Microbiol. Serol.*, 46, 595, 1980.
136. **Yamada, Y., Sasaki, J., Hou, C. F., Tahara, Y., Miranda, M., and Phaff, H. J.,** A new natural quinone, ubiquinone-5 isolated from an ascosporogenous yeast species *Nematospora coryli*, *Agric. Biol. Chem.*, 45, 2671, 1981.
137. **Aulakh, H. S., Straus, S. E., and Kwon-Chung, K. J.,** Genetic relatedness of *Filobasidiella neoformans (Cryptococcus neoformans)* and *Filobasidiella bacillispora (Cryptococcus bacillisporus)* as determined by deoxyribonucleic acid base composition and sequence homology studies, *Int. J. Syst. Bacteriol.*, 31, 97, 1981.
138. **Phaff, H. J., Miller, M. W., and Miranda, M.,** *Hansenula alni*, a new heterothallic species of yeast from exudates of alder trees, *Int. J. Syst. Bacteriol.*, 29, 60, 1979.

Chapter 4

THE PROBLEMS OF LARGE-SCALE YEAST PRODUCTION

G. E. Guidoboni

TABLE OF CONTENTS

I. INTRODUCTION

During the 1950s various studies predicted that during the second half of the 20th century the classical sources of protein production would not be able to produce the quantities needed. Total world protein production was running at an annual level of over 80 million tonnes, against a corresponding world deficit, at the time, estimated at nearly 25 million tonnes. Increased protein production from the sea was considered, but the sort of increase called for was beyond any reasonable expectation (all the world catch of fish only accounts for about 3% of the world current protein consumption). The other obvious source, the land, was considered, but it was recognized that while cattle grazing could have been increased, it would have been achieved at the expense of cereal production, since farmable land was a limited resource. Additionally, much research was being carried out which was devoted to seeking out new unconventional sources of protein with work concentrated on growing seaweed and producing microorganisms. The latter, in the form of yeasts and bacteria, offered the advantage, apart from their short growth period, of being able to convert non-organic and, therefore, inexpensive forms of nitrogen into high quality protein.

In the 1950s and early 1960s there was a burst of activity in this field of protein from unconventional sources and much interest was aroused in the processes being developed by British Petroleum for yeast from hydrocarbons. The original process consisted of growing yeast on gas oil and the second and complementary process used preselected normal paraffins which were entirely consumed by the yeast. Development of the normal paraffin process was carried out through the 1960s and 1970s and included the building and operation of a 4000-tpa commercial demonstration plant. By 1976 a full-scale 100,000-tpa production plant had been built and was ready for operation, before the oil crises (rapidly escalating feedstock costs) and politics put the commercial viability of the project into jeopardy. (The subsequent events leading to the abandonment of the project are now history.)

The problems encountered in the development of the normal paraffin-based process and its operation on a large scale form the basis of this paper. The experiences gained may be applied to other large-scale yeast production processes where the yeast may be the end product in itself (single-cell protein) or may be the vehicle for the large-scale production of other materials.

II. PROCESS BASIS

The process outlined below forms the basis of the discussion for this paper and involves the fermentation of a hydrocarbon feedstock in truly continuous operation under strict monoseptic conditions.

A schematic flowsheet is shown in Figure 1 and the basic elements of the process can be seen. In this process the fermenter operates under closed conditions so that one single-type organism exists in the fermenter vessel and close control is maintained by careful manipulation of pH, temperature, etc.

The raw materials are fed continuously into the fermenter via sterilization equipment. The liquid feedstocks are sterilized by heating and holding at a high temperature for a sufficient length of time to ensure a complete "kill" of all microorganisms that may be present in any of the feed streams. Heat economy is achieved by exchange between the incoming cold feed and hot, sterilized feed.

All the oxygen necessary for cell growth is supplied by a continuous stream of air blown into the fermenter and the nitrogen, essential for cell growth, is supplied by a flow of ammonia gas which is fed continuously into the air stream. The ammonia also serves to maintain the required pH of the fermenter contents and the rate of ammonia addition is automatically controlled.

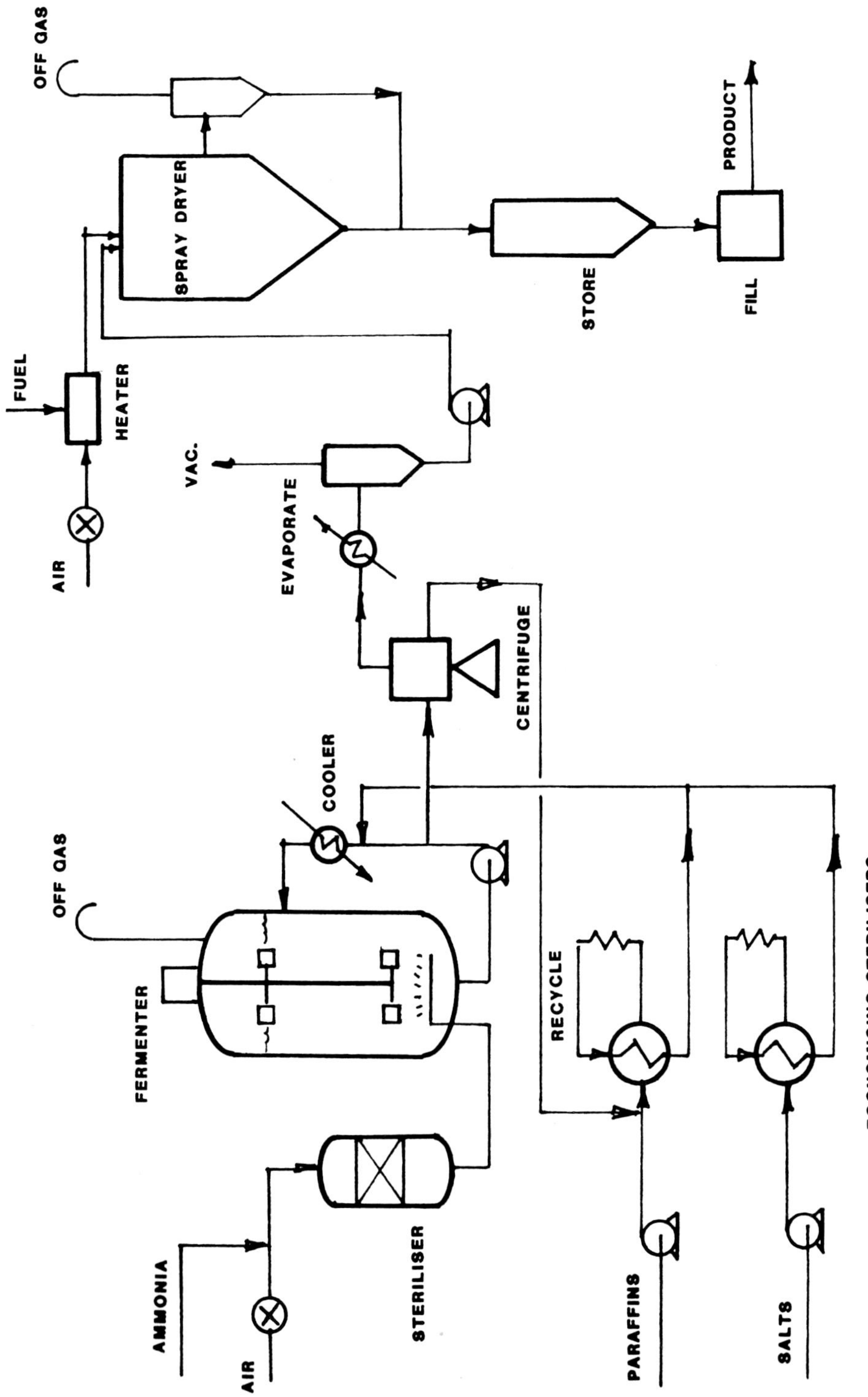

FIGURE 1. Schematic flowsheet.

The mixed air and ammonia gas stream is sterilized by filtration and the off-gas from the fermenter passes directly to atmosphere via a vent stack.

Within the fermenter vessel, transfer of hydrocarbon, essential minerals, and oxygen to the cells is achieved by intimate mixing of the four phases by means of mechanical agitation. The high metabolic heat release is removed by heat exchange in an external cooling loop and the temperature within the fermenter is controlled by regulation of the flow of cooling water to the external heat exchanger.

The product from the fermenter passes through a yeast/water centrifuge separator stage to give a cream product of about 15% wt solids and clear discard aqueous phase. A further concentration of the cream product, up to about 25% wt solids, takes place in a multistage evaporator before the yeast cream passes into the final drying stage.

From the drying stage, the powder product is pneumatically conveyed, cooled, and discharged into storage hoppers. The product is bagged directly from these hoppers for dispatch.

III. SOME ANTICIPATED SCALE-UP PROBLEMS

A. General

Early economic studies showed quite clearly that production of protein concentrate from hydrocarbons would need to be at higher annual tonnages per plant than those normally achieved by traditional yeast fermentation plants. Operation, therefore, would have to be truly continuous and the individual section of the proposed plants would have to employ either the largest available standard item of equipment or multiples of it. Hence, the design of even a relatively small 4000-tpa commercial-scale protein plant presented many problems, and it was realized that in many areas the techniques used even in the largest of pilot plant fermenters and harvesting schemes would not be applicable directly to full-scale operation. Also, because at the time the design for the 4000-tpa plant was begun, there did not exist a body of expertise in equipment manufacture for the handling, processing, etc. of the type of yeast used in this particular process. However, there was a wide background of information on the handling of bakers and brewers yeast, but experience soon showed that the hydrocarbon fermenting yeast was quite unlike the other yeasts in many respects.

B. Feed Preparation

Because of the volume of liquid feeds involved in the 4000-tpa process, the continuous sterilization of those feed streams presented a problem as to what method of sterilization would be practicable and how would such a system be arranged. Various means of sterilization had been considered in theory and several had been tried in practice in both the laboratories and pilot plants. If we consider, for instance, heat treatment, a fairly conventional exercise in heat transfer design was carried out to achieve both maximum heat economy and minimum heat transfer area. This second point was vital because of the high cost of the materials in the heat exchanger and associated pipework needed to withstand the severe corrosion attack of the input medium. The piping design of the liquid sterilization system was greatly complicated by the need to ensure that during start-up, or change of feed rate to the fermenter, any material fed forward would be completely sterilized at every instant of time.

It was anticipated that recycle of aqueous streams from the centrifuges to the fermenters would lead to fouling of the sterilizing exchangers and a cleaning-in-place technique was developed to control this particular problem.

The problems involved in sterilizing the very large volume flow of air and ammonia were considerable, and after several years of equipment testing it was realized that proprietary equipment was unsuitable for various reasons. Pressure drops were unacceptably high in some units, and other units could not withstand initial steam sterilization at start-up.

The units which were available commercially at the time consisted of a multitude of individual filter elements fitted into a base plate and housed inside a common shell. It was considered that with the large number of joints involved the chances of satisfactory operation were slim. Accordingly, development work was carried out with various filtration media in order to design an air sterilization filter that could be built as one unit. By consideration of the X_{90} values for various media, glass fiber of a certain mean fiber diameter was selected. The medium chosen has an X_{90} value of a few centimeters and a reduction in organism count between inlet and outlet gas streams of $>10^{10}$ was achieved.

C. Fermentation

The heart of the whole process is, of course, the fermentation stage and it is here that the greatest problems arise. Basically, the main problem arises from the need to achieve proper contacting of four separate phases within the fermenter, viz. the two immiscible liquid phases (water and hydrocarbon), the solid phase (the yeast), and the gas phase (the air/ammonia mixture). Many years of experimental work had been carried out in the pilot plant and the scale-up calculations for the first industrial-scale fermenter vessel were based on the results obtained from agitation and mass transfer studies in pilot units with operating volumes separated by the ratios 1:10:100. The way in which operating data from the commercial demonstration plant were used in conjunction with pilot plant data to establish further reliable scale-up correlations will be discussed later.

D. Product Recovery

Because most of the process development work over the years had concentrated on the fermentation stage, the harvesting train for the first commercial plant was based on established techniques for baker's and brewer's yeast recovery, while being aware of possible complications mentioned earlier by virtue of being first in the field.

The basic problem in biomass harvesting from *n*-paraffins fermentation is the recovery of a relatively small amount of solid yeast from an enormous quantity of water (quite the opposite to most other fermentation processes where the desired product is in the aqueous phase). Therefore, there is a major solid/liquid separation problem, complicated by the need for a dry product.

Centrifuges have been used for a long time for the recovery of solids from liquids, and the problem was not one of principle, but of prediction of separating efficiency and machine throughput capability when handling a new material. It was realized that a considerable amount of experimental work on a commercial scale would be needed.

In the area of evaporation of protein cream, the problems in design of the commercial unit centered around the choice of heat transfer coefficients and hydraulic loading to prevent protein fouling of the heat transfer surface. Design of the commercial unit was based on prior operating experience with other types of yeast creams and data obtained from pilot-scale experiments with proprietary test models, but it was accepted that full-scale evaluation trials would have to be carried out before any very large units could be designed with confidence.

For the first demonstration plant a spray dryer was selected for the final drying stage and the design of the unit was straightforward. The only problem lay in the prediction of the particle size and particle size distribution of final product expected in the large commercial unit.

The design of the storage and bagging facilities did not appear to pose any major problems, and equipment was selected on the basis of results obtained from trials with pilot-scale-produced protein powder.

IV. OPERATIONAL PROBLEMS

The previous section describes some of the scale-up and other problems which were anticipated and some indications given as to how the problems might be tackled. Some of the actual problems encountered are now described, together with solutions.

A. Commercial Demonstration Plant

The plant referred to here was a 4000-tpa protein production unit which was built to prove the overall process at a commercial scale and to provide a finished product for market evaluation and cultivation.

Initially, trouble was experienced in sustaining an aseptic fermenter operation. By attention to detail in the initial sterilization of equipment, and the continuous sterilization of the feed streams, satisfactory operation was achieved. The number of piping flanges was reduced where possible. Pipe runs were improved in order to remove static pools of liquid that could render the initial sterilization procedure ineffective. The insulation on the fermenter section was improved to ensure that all parts achieved a minimum temperature during steaming.

Plate heat exchangers were used to achieve heat economy in the sterilization of the liquid medium. Problems were experienced with the materials of construction, both metal and the jointing material, to withstand the temperature with an acid solution containing hydrocarbons, sulfate, and chloride ions.

To control the residual hydrocarbon and ash content, it was necessary to have rigid control of the hydrocarbon and ion addition rates. By installing the correct pumps and measuring tanks to check flow rates, satisfactory control was achieved. Surprisingly, the fermenter on the demonstration unit, which was larger in volume by a factor of 100 than the pilot fermenter from which the basic design data were obtained, gave few problems. It achieved the desired productivity from the moment that operation was sufficiently smooth to allow an assessment of this parameter. Some relatively minor mechanical problems arose with the agitator, but these were rapidly corrected. Foaming was not a significant problem. On occasions when there was foam at the broth surface, it was adequately handled by the upper impeller. The air compressor gave virtually trouble-free operation.

The centrifuges installed to harvest the biomass as a 15% wt cream performed well. The major problem was one of hygiene, in that regular cleaning of the machines was necessary in order to produce a product of excellent microbiological quality. A cleaning-in-place procedure was developed to avoid the necessity for stripping down the machines.

It was originally intended that the cream would be concentrated from 15 to 25% wt by evaporation, prior to spray drying. However, because of unsteady plant operation upstream, sustained satisfactory operation of the evaporators was not obtained, and the 15% wt cream was fed direct to the spray dryer.

The spray dryer performed adequately, but as with the centrifuges, it was necessary to operate a cleaning schedule in order to ensure a product of satisfactory microbiological quality.

V. DESIGN PROCEDURES

As mentioned in a previous section, the fermentation is the heart of the process and it is the stage which was studied most intensively over the years by the development team. The main aim for chemical engineers was the development of correlations for the scale-up and design of large industrial continuous fermenters, and some of the ways in which this problem was attacked are discussed.

Basically, there are three stages in the design of large fermenters for which scale-up and design correlations are needed, viz:

1. Oxygen requirement for cell growth
2. Prediction of oxygen transfer rate
3. Prediction of power dissipation to achieve the required oxygen transfer rate

A. Oxygen Requirement for Cell Growth

Over the years many oxygen balances have been carried out on fermenters of various sizes, from 20 ℓ to 200 m^3 working volume, and the oxygen requirements for cell production have been established. The results have compared well with values predicted by equations such as the one below:

$$2nCH_2 + 2nO_2 + 0.19NH_4^+ + (P, K, S, etc)$$

$$n(C_1H_{1.7}O_{0.05}N_{0.19}\ Ash) + nCO_2 + 1.5nH_2O + 824n\ MJ\ (Ref.\ 1)$$

This particular equation shows that the oxygen requirement for cell growth, in terms of grams oxygen per gram of cells, is related to the number of carbon, hydrogen, and oxygen atoms in each molecule of carbon source, the number of carbon, hydrogen, and oxygen atoms in the cells, the cell yield factor, and the molecular weight of the carbon source. The agreement between the calculated and measured oxygen requirement was sufficiently close for large-scale plant design purposes and most of the development work was concentrated on the remaining two topics.

B. Oxygen Transfer Rate

For large-scale plant design a correlation is needed which will enable the yeast production rate to be predicted from the fermenter geometry and operating conditions. Under the conditions in this hydrocarbon process the fermenter performance is controlled by oxygen transfer from the air bubbles to the cells. This has been shown to be related to the yeast production rate by the equation:

$$P = x\ Y_o\ R_o\ V$$

where P = yeast production rate kg/sec, x = constant, Y_o = kg yeast produced per kg of oxygen, R_o = oxygen transfer rate kg · mol/m^3/sec, and V = ungassed liquid volume m^3.

If the system is oxygen limited it is necessary to be able to calculate the oxygen transfer rate for fermenter design purposes. There were several relationships available from the literature,[2-4] but because some of the relationships were based on sulfite oxidation experiments and because the mass transfer across the stagnant liquid film at the gas-liquid interface is affected by the presence of surfactants, electrolytes, proteins, and solid particles (all of which are present in the hydrocarbon fermentation), it was likely that the oxygen transfer rates predicted by these equations would differ from those existing in this particular fermentation system. Hence, it was necessary to use experimental data from the available fermentation systems to determine the numerical values of the various constants in any of the possible equations. Of the correlations examined, the semitheoretical relationship proposed by Westerterp et al.[2] was selected for more detailed examination. Experimental data were obtained from fermentations carried out on pilot scale (maximum working volume of about 2 m^3) and were used together with data generated on the commercial demonstration plant to examine the applicability of the Westerterp correlation to this particular fermentation. When using Westerterp's numerical values for the constants in the equation given below, a relatively poor fit was obtained.

The original equations were

$$Ro = \frac{\phi in \; Xin}{22.4V + \dfrac{\phi out \; R\theta}{W \; \pi T}}$$

$$\text{where } W = \alpha\left\{\frac{DT^{1/2}}{H}\right\}(n\text{-}n_o)$$

$$\text{and } \quad n_oD = \beta\left[B + \left(\frac{CT}{D}\right)\right]$$

where B and C are constants, D = impeller diameter m, H = ungassed liquid height m, R = gas constant, Ro = oxygen transfer rate kg · mol/m^3/sec, T = vessel diameter m, V = ungassed liquid volume m^3, n = stirrer speed S^{-1}, n_o = critical stirrer speed S^{-1}, X in = volumetric fraction of oxygen in inlet gas, ϕ in = inlet gas rate m^3/sec, ϕ out = outlet gas rate m^3/sec, θ = absolute fermentation temperature °K, T = top pressure kN/m^2, and α, β = constants.

Therefore, the basic equations were modified and new values of constants were obtained by regression analysis of experimental data from pilot and commercial scales. The new fit of experimental vs. measured oxygen transfer rates was very much more satisfactory and adequate for design purposes.

C. Power Dissipation

Any equations for power consumption must relate to the actual power input to the fluid contents of the fermenter. The power input to the motor drive system must be corrected for the power losses in the system. In general, these losses are a function of the agitator speed and the power transmitted. It is considered that the power actually transmitted to the fermenter broth should be used to calculate the specific power consumption (kW/kg dry protein) and not the power input to the motor. This is particularly true of small fermenters whose agitator drive systems usually have lower efficiencies than the much larger industrial systems. The actual power consumption for any given system can then be calculated from the specific power consumption using the efficiency of the drive system.

The agitator power consumption problem may be considered in two parts. First, there is the power consumption of an agitator in a homogenous Newtonian liquid, and second, there is the power consumption in a gassed liquid.

The equations used to represent the agitator power consumption in a homogenous liquid under turbulence conditions are well established and take the form:

$$P = k\rho \; N^3D^5$$

where P = power transmitted to the liquid, k = constant, ρ = liquid density, N = stirrer speed, and D = stirrer diameter, all in consistent units.

The constant, k, is a function of the system geometry and impeller type and can be determined for each system.

However, the estimation of the power consumption of an aerated and agitated system is less satisfactory, and of the various equations published, the one due to Michel and Miller[5] was examined for applicability to the hydrocarbon fermenter system. This equation has the form:

$$P_g = a\left\{\frac{Po^2 \; ND^3}{Q_b}\right\}^c$$

where a, b, and c are constants, P_g = power transmitted to aerated liquid, Q = air rate, and P_o = power transmitted to ungassed liquid (other symbols as in previous equation).

This equation was somewhat unsatisfactory because it did not take into account the ratio of impeller-to-fermenter diameter (their final equation did not contain D/T as a variable although it showed separate curves for various D/T ratios) and other aspects of vessel geometry. Since the Michel and Miller paper in 1962 various workers have modified the original equations slightly to allow for such effects as D/T ratio and vessel volume on the value of the constants a and c.

Using data obtained from pilot-scale fermenters and the commercial demonstration unit, it was possible to modify the basic equations still further and to determine the value of the constants in the equations. The equations developed took into account the effect of D/T ratio and the vessel geometry.

This modified equation has the form

$$P_g = a\left\{\frac{Po^2\ ND^3}{Q^b}\right\}^c \left[\frac{x}{D}\right]^d$$

where x is the height of the lower impeller above the sparger and d is another constant.

Although the modified equations for oxygen transfer rate and power dissipation covered satisfactorily a range of fermenter sizes in the ratio of approximately 125:1, experimental work was carried at various scales of operation to refine these equations and allow them to be used for an even greater degree of scale-up. Such equations were used in the basis of design for the fermenters for protein plants of 100,000 tpa capacity.

D. Other Design Matters

Many new and complex problems were encountered in developing the process from its 4000-tpa demonstration scale to the 100,000-tpa production scale. It was necessary to sterilize nearly 150 t/hr of fermenter feed streams, while circulating many times this flow through each broth cooling circuit. Some areas of the plant required the development of equipment on an unprecedented scale, which also had to operate under aseptic or sterile conditions.

1. Development of Agitators and Fermenters

The fermenter/agitator system for the plant was the largest in the world. Agitator-consumed power was about 2500 kW and broth volume was over 1200 m^3. In addition to coordinating activities of the agitator and vessel vendors, engineers undertook the detailed computerized vibrational analysis of the system. The agitator shaft seal was designed to provide a sterile barrier despite shaft vibration, flexing, and expansion.

In mechanically agitated vessels, mechanical design problems can arise from vibrations caused by the rotation of the agitator and the movement of the liquid. Measurements were made at a number of liquid levels, air rates, f and agitator speeds on the 4000-tpa plant using accelerometers and strain gauges. Some of the measuring devices were attached to shaft and paddles.

The data obtained were analyzed by computer to give a model which has been employed in the design of the 100,000-tpa fermenter vessels, agitators, foundations, and supporting structures. Data of this nature are not available from agitator manufacturers.

2. Broth Circulation Pumps

Each pump had to circulate broth (7000 t/hr) containing 15% by volume of air. To pump these quantities of broth under aseptic conditions required the development of special pumps. The development team collaborated closely with the vendor to develop these pumps.

3. Reduction of Heat Load

The fermenter feed streams represent a significant proportion of the fermenter cooling load. For most of these streams further cooling offers no potential economy, however, studies indicated that significant savings could be achieved by incorporating an air cooler in the air line to the fermenter. This feature was incorporated in the final design.

4. Cleaning-in-Place

The requirements for the sophisticated cleaning-in-place system were developed from first principles. A computer-controlled system was selected on account of cost, versatility, and data recording ability.

5. Effluent Treatment

The original effluent treatment scheme proposed for the plant was extremely complex. Two alternatives were examined:

- A simplified but efficient scheme to reduce costs
- Elimination of the effluent treatment plant and recycle of all supernatant from the centrifuges to the fermenter feed system

After tests at the commercial demonstration plant, the latter system was adopted and a considerable cost saving achieved. A small effluent treatment plant was still required to treat spillages and drainings.

6. Energy Savings

A number of studies into the use of alternative machine drives and other cost reduction techniques were carried out. The most important results of such studies incorporated in the final design were

- Optimization of broth cooling circuit capital and running costs
- Use of exhaust gases from gas turbines as hot air supply for product drying

VI. CONCLUSIONS

The development of a truly continuous monoseptic yeast fermentation process at a large scale presented problems, to both microbiologists and chemical/biochemical engineers, that had previously not been experienced. The technical success of the development program was achieved in the face of considerable scepticism, both academic and industrial, and a number of lessons were learned the hard way.

- A body of background information and knowledge does not exist from which the process designer can draw; where it does exist it is generally commercially secret.
- An extremely high degree of cooperation is needed between the microbiologist and chemical/biochemical engineer in order that their unique individual problems may be jointly resolved.
- Monoseptic fermentation can be achieved in the continuous mode in mechanically agitated fermenters up to 1250 m^3, but only by rigorous attention to minute engineering detail.
- Downstream processing of living microorganisms presents as many problems and challenges as fermentation, especially where the microbiological quality of the product is concerned.
- Truly continuous operation at a large scale generates unprecedented problems for plant

and equipment fabricators, at times way beyond those encountered even in the best pharmaceutical fermentation processes.

- Pilot testing of proprietary equipment is an absolute necessity; reliance cannot be placed on prior experience with "so-called" similar products.
- In spite of all the potentially overwhelming problems, "IT CAN BE DONE."

REFERENCES

1. **Bennet, I. C., Hondermarck, J. C., and Tood, J. R.,** How BP makes food protein from hydrocarbons, *Hydrocarbon Process.*, 48, 3, 1969.
2. **Westerterp, K. R., van Dierendonck, C. L., and de Kraa, J. A.,** Inter-facial areas in agitated gas-liquid contactors, *Chem. Eng. Sci.*, 18, 157, 1963.
3. **Calderbank, P. H. and Moo-Young, M. B.,** The continuous phase heat and mass transfer properties of dispersions, *Chem. Eng. Sci.*, 16, 39, 1961.
4. **Yoshida, F. and Miura, Y.,** Gas adsorption in gas-liquid contactors, *Ind. Eng. Chem. Proc. Des. Dev.*, 2, 263, 1963.
5. **Michel, B. J. and Miller, S. A.,** Power requirements of gas-liquid agitated systems, *AIChE J.*, 8, 262, 1962.

Chapter 5

YEAST BREEDING FOR FUEL PRODUCTION

Flavio C. A. Tavares and Sergio Echeverrigaray

TABLE OF CONTENTS

I. SUMMARY

Higher yields and productivities are the main objectives of yeast breeding for fuel alcohol production. Polygenic characters related to fermentation require proper breeding strategies that must follow any transfer of major genes to prepare the strain for industry.

Results are presented of a breeding program developed for batch and continuous fermentation of sugar cane by *Saccharomyces* sp., fermentation of inulin from Jerusalem artichokes, and starch. Recombinational methods were applied and accordingly natural crosses, protoplast fusions, and rare matings were used for construction of hybrids. Special breeding methods including Hybridization Breeding, Recurrent Selection, Polycross Breeding, and Selection after Self-Crossings have provided strains for ethanol production in Brazilian industries. Promising strains selected on a laboratory scale were tested in scale-up assays and one of the hybrids has been used in industry in recent years. Fermentation with such hybrid is estimated to account for about 10% of the Brazilian ethanol production. Data from industry are in good agreement with yields and productivities obtained in bench-scale experiments. Strains of *Kluyveromyces marxianus* have been selected and their genetics are now the object of attention. Fermentations with such strains indicate promising possibilities of this kind of ethanol production from the enzymatic hydrolysis of inulin. The best results for starch fermentation after fusions among different yeasts are presented by some *S. diastaticus* hybrids; in addition, cloning experiments with *Schwanniomyces castellii* are presently underway. Single genetic factors for flocculence and killer have been transferred to industrial strains. A tower fermentation using flocculent yeast has been developed and killer yeasts, after industrial tests, were found to be very efficient.

II. INTRODUCTION

The use of ethanol from biomass as a liquid fuel has been successful in the past, although production has been limited due to competition from petroleum products. However, after the oil crisis of the 1970s, ethanol became an important alternative fuel and Brazil demonstrated its viability in large-scale production. Through governmental action, the Alcohol National Plan (PNA) started in 1975, with the aim of supporting growth of the ethanol industry. Since then, production has been rising and it is presently over ten million cubic meters, with the potential to produce even more for export.

In Brazil such large-scale production is dependent on yeast fermentation of sucrose from sugar cane and, to a lesser extent, starch from cassava after hydrolysis. Most yeasts currently used are of the baker's type. In general, industrial efficiency would be improved if additional process controls and more adequate yeasts were used. The yeasts, for instance, should have the potential to assure high product yields and productivities related to many known polygenic characteristics. Abilities such as vigor, thermotolerance, alcohol tolerance, tolerance to inhibitor substances (e.g., organic acidity), and fermentation rate are among the characteristics needing improvement. It is also important to adapt yeast strains to different technological processes, including specific equipment limitations.

These considerations necessitated the initiation of a program for breeding superior strains of yeast for ethanol production. The results of this program are presented in detail below.

III. YEAST PROPERTIES FOR FUEL ETHANOL PRODUCTION AND RECOMBINATION METHODS FOR YEAST BREEDING

Despite efforts to maintain a regular fermentation process, it is almost impossible to develop uniform conditions for a given growing season. To do so requires breeding yeast able to retain its fermentation properties over a broad variety of environments in order to

Table 1
PRODUCT YIELD REGRESSION ANALYSIS AND CORRELATION VALUES WITH RESPECTIVE F TEST SIGNIFICANCE

		Regression equation		
Parameters	**R**	**Linear**	**Quadratic**	**Cubic**
Acidity (g/ℓ)	−0.186	*		
Temperature (°C)	0.126		**	
Ethanol (%)	0.228		**	
Time (hr)	−0.267			**
Batches (N°)	−0.131		**	
Cell mass (%)	0.108		**	

maintain elevated and consistent rates of production. Some biological properties such as osmotic pressure and temperature tolerance,[1-8] which are of particular interest to our program, have been studied. In order to evaluate the selection criteria adopted and to obtain additional information on how particular industrial parameters relate to the biological properties of a yeast, regression analysis of about 70 variables was made for every crop season. Specific correlations between characteristics were obtained and although a single characteristic could not be identified, some of the generalized traits do show early repetition. There is a significant correlation between sugar cane quality and the preparation of the worts to be fermented, related to product yield. For instance, if taken alone, the acidity of the worts as an indicator of quality shows a negative correlation with product yield (Table 1) and its physiological effect can be manifested by inhibiting budding and decreasing cell viability. To some extent tolerance to acidity is a desired property, as is also observed for temperature and ethanol content in wines which have optimum ranges for high yield values. For the yeast hybrids analyzed, the optimum temperature was 37°C and (for the crop season mean) the alcohol content 9°GL, with volumetric and specific rates of 8.54 $g/\ell \cdot h^{-1}$ and 0.35 $g/g \cdot h^{-1}$, respectively.

Other variables such as fermentation time or batches processed per day revealed a negative correlation with product yield, an indication that productivity and product yields cannot be dissociated. Also, due to the nature of the regression equation there are limits where high yields and maximum productivity occur. Cell viability and budding, as well as cell concentration, usually present a positive correlation, indicating a tolerance to inhibitory factors and an ability to rapidly multiply in order to compensate for cell loss during the process.

Until the development of techniques for the direct transfer of genetic material among different microbial species, mutation and selection were the universal methods adapted for strain improvement. Although still important today, mutation has, in many instances, been substituted for genetic recombinational methods.[9] This idea is not new, in fact Lindegren,[10] in 1949, pointed out that yeast hybridization was of potential interest and proposed a method to obtain strains by "rare mating" able to grow in highly unfavorable conditions, such as sulfite liquor. He suggested that interspecific hybridization could be employed in a breeding program for alcohol production. Quantitative studies in microorganisms are the basis for successful breeding and understanding continuous variation.[11-17] The breakdown of species barriers by genetic recombination has allowed for the manipulation of polygenic characters to develop novel strains of yeasts. These, coupled with the use of natural variability in combinations with other methods, provide the microbiologist with extremely powerful tools for breeding novel yeasts.

Results have been previously presented[18,19] showing the utility of adopting plant breeding methods for yeast improvement for ethanol production. As has been observed in other cases,

Table 2
ETHANOL PRODUCTION AND PRODUCT YIELDS IN SELECTED STRAINS OF YEAST COMPARED TO INDUSTRIAL TESTERS M 300 A AND FLEISCHMANN YEASTS

Strains	Ethanol (g/100 mℓ)	Product yields (%)
M 300 A	8.446 ± 0.886	93.36
FF	8.280 ± 0.860	91.52
IZ 1904	6.932 ± 5.296	76.62
IZ 608	8.194 ± 1.383	90.57
IZ 862	7.860 ± 0.506	86.88
TC_1	8.412 ± 0.549	92.98
TC_2	7.208 ± 1.383	79.67
E 202	8.596 ± 0.852	95.02

polygenic characteristics related to fermentation require proper breeding strategies that must follow the transfer of major genes to adapt yeast strains for industrial usage. Some breeding methods have been developed in order to select superior recombinants that are better fermenters.

In industrial ethanol production relatively little attention has been given to the choice of which yeast to employ, since commercially available baker's yeast offers the facility of a large inocula for fermentation. In some distilleries, however, where operational difficulties occur or where high productivities and yields are the goal, more specific yeasts are generally selected. As is well known, productivity (general and specific) depends on cell mass concentration, velocity of product formation, and on cellular efficiency at maximum product formation,[20] and considering the complexity of the characteristics involved in productivity and yield, the selection criteria and adequate evaluation of the biological material have made breeding a complicated task. Some breeders have worked to improve very specific characteristics; however, considering that both ethanol productivity and product yields are to be high and associated in the selected yeasts, various biological characteristics must be manipulated. This fact demands the development of new selection and evaluation methods.

Through the standardization of experimental conditions and acceptance of some characteristics as basic for alcohol production, we have been able to select superior strains from a diverse collection of yeasts. Hundreds of strains were tested and compared to several common industrial strains under a standard sugar concentration (20%), temperature of incubation (35°C), and consecutive fermentations, each of them spaced by a daily period of starvation at low temperature. Relative production of ethanol as compared to the tester strain ranged from 9.21 to 145.33%. As shown in Table 2, yields for several of the experimental strains were superior to at least one of the industrial strains. Other parameters, derived after evaluation of alcohol production kinetics, show the qualities of the selected strains (Table 3). The effect of temperature and osmotic pressure on yield can be observed in Table 4 where reduction of production was almost 50% for every 5°C increase in temperature.

Evaluation of the results of strain selection from diverse origins showed that the use of superior strains for hybridization did not produce the best hybrids or segregant strains. This result has necessitated a simultaneous selection scheme after crossing for the production of segregants and new hybrids. Such segregant selection has produced a number of superior strains, some of them with higher means than the tester strains and small standard deviations. The fermentation kinetics of these strains were evaluated (Table 5). As can be observed, fermentation rates are not affected by efficiency of substrate conversion to ethanol. This effect can be associated with inhibition by ethanol as observed previously.

Table 3
KINETICS OF ALCOHOL PRODUCTION FOR SELECTED YEAST STRAINS

	Strains				
Hours	**IZ 608**	**IZ 862**	**TC_1**	**M 300 A**	**E 202**
1	1.24	0.99	1.34	1.67	1.52
2	2.13	1.85	2.38	2.75	2.49
3	3.14	2.46	3.32	3.79	4.12
4	4.71	4.05	4.57	5.13	5.49
5	5.71	4.97	5.73	6.23	6.35
6	6.35	5.53	6.28	6.96	6.99
7	7.34	6.91	6.58	7.61	7.64
8	7.45	6.63	7.33	7.99	8.07
22	10.06	9.94	9.87	8.58	8.90

Note: Ethanol (g/100 mℓ) production values are average of three consecutive fermentations.

Table 4
THE EFFECT OF TEMPERATURE AND OSMOTIC PRESSURE (15—25% SUGAR) ON ETHANOL PRODUCTION (E, g/100 mℓ) AND YIELDS (R, %) OF SELECTED STRAINS

	Reducing sugar conc							
	35°C — 20%		**35°C — 25%**		**40°C — 20%**		**40°C — 25%**	
Strains	**E**	**R**	**E**	**R**	**E**	**R**	**E**	**R**
M 300 A	10.2	(80)	10.2	(64)	6.1	(48)	5.0	(31)
FF	9.8	(76)	8.6	(54)	4.6	(36)	4.3	(27)
IZ 608	10.3	(80)	10.7	(67)	4.8	(38)	5.6	(35)
ATCC 4126	10.0	(78)	9.7	(61)	4.7	(37)	3.3	(21)
J 132 b	9.3	(73)	8.9	(59)	5.7	(45)	4.5	(28)
MONTRACHET	10.1	(79)	10.9	(68)	5.0	(39)	5.1	(32)
	35°C — 15%		**35°C — 20%**		**40°C — 15%**		**40°C — 20%**	
	E	**R**	**E**	**R**	**E**	**R**	**E**	**R**
M 300 A	6.6	(65)	7.6	(59)	5.5	(54)	2.0	(15)
FF	6.9	(68)	7.8	(60)	4.9	(48)	2.8	(22)
F1-A	6.9	(68)	6.5	(50)	3.5	(34)	2.5	(19)
F1-C	6.9	(68)	7.8	(60)	4.0	(39)	2.2	(17)
IZ 1347	7.0	(69)	7.7	(60)	4.3	(42)	3.6	(28)
ATCC 26603	6.4	(63)	6.3	(49)	4.9	(48)	3.9	(31)
	6.9	(68)	7.2	(56)	4.8	(47)	1.6	(12)
	35°C — 20%		**35°C — 25%**		**40°C — 20%**		**40°C — 25%**	
	E	**R**	**E**	**R**	**E**	**R**	**E**	**R**
M 300 A	9.9	(71)	8.6	(51)	5.5	(39)	3.4	(20)
FF	8.8	(63)	7.3	(43)	4.4	(31)	2.4	(14)
E 202-IV-39	10.1	(72)	8.0	(47)	5.8	(41)	4.3	(25)
M 304-2C	11.1	(79)	8.2	(48)	7.1	(51)	3.7	(22)

Table 5
KINETICS OF FERMENTATION USING SELECTED STRAINS AND SEGREGANTS

	Strains						
Hours	M 300A	FF	1Z 1904	M 2-A	F 9-B	F 1-C	F 6-C
1	0.19 ± 0.07	0.20 ± 0.11	0.21 ± 0.18	0.24 ± 0.05	0.27 ± 0.02	0.23 ± 0.002	0.21 ± 0.04
2	0.63 ± 0.21	0.69 ± 0.28	0.68 ± 0.27	0.79 ± 0.08	0.78 ± 0.24	0.64 ± 0.37	0.65 ± 0.25
3	1.37 ± 0.14	1.47 ± 0.20	1.37 ± 0.38	1.65 ± 0.24	1.72 ± 0.21	1.67 ± 0.41	1.51 ± 0.07
4	2.08 ± 0.13	2.20 ± 0.06	1.99 ± 0.56	3.37 ± 0.06	4.17 ± 0.53	2.50 ± 0.25	2.26 ± 0.20
5	3.03 ± 0.46	3.15 ± 0.51	2.79 ± 0.67	4.74 ± 0.50	3.60 ± 1.20	4.29 ± 2.11	3.19 ± 0.67
6	3.70 ± 0.23	3.82 ± 0.32	3.33 ± 0.71	4.35 ± 0.27	4.27 ± 0.80	4.11 ± 0.47	3.75 ± 0.48
7	4.39 ± 0.24	4.24 ± 0.24	3.52 ± 0.32	5.13 ± 0.15	4.92 ± 0.89	4.92 ± 0.08	4.42 ± 0.50
8	4.99 ± 0.28	5.15 ± 0.23	3.93 ± 0.36	5.79 ± 0.20	5.40 ± 1.00	5.48 ± 0.02	4.88 ± 0.56
15	9.12 ± 0.18	9.50 ± 0.06	7.73 ± 1.58	10.12 ± 0.91	9.61 ± 0.88	9.15 ± 0.81	8.81 ± 0.03

Note: Alcohol values are in g/100 mℓ, average of three fermentation runs. The initial reducing sugar concentration was 20% w/v and temperature of fermentation was 35°C.

Hybrids have been obtained, and were selected and tested simultaneously against respective parental strains. These have permitted the identification of the best hybrids which in turn were also evaluated for fermentation kinetics (Table 6). Production yields (g/100 mℓ) were significantly higher than the industrial tester strain or fairly close to the best segregant obtained (M 304-2C). Maximum alcohol concentration occurring after 15 hr was obtained after total sugar conversion. As can be seen in Table 7, volumetric productivity was almost 50% higher than for the industrial strain M 300A. These are higher values than reported before for discontinuous fermentation, and are close to production values obtained by continuous fermentation.[21] Temperature and osmotic pressure effects on selected hybrids were evaluated and these results appear in Table 8. The stressing effects of both factors are pronounced mainly at 40°C and 25% total reducing sugar.

As mentioned, the segregant strain M 304-2C exhibited interesting properties that could be exploited by recurrent selection. In order to complement the evaluation of this strain, an experiment was devised involving a series of discontinuous fermentations, with and without aeration, varying fermentation times, and exposure to the toxic effect of ethanol and sugar starvation. A summary of these data appears in Table 9. Alcohol production and yields were high. In continuous fermentation this strain, which was stable for 33 hr, had maximum ethanol production of 102.5 g/ℓ. The overall productivity reached 21.53 g/ℓ · hr, with a production yield of 93.2% observed. An experiment using rapid fermentation[1] was conducted with 35% sugar worts and a cell content of 117 g/ℓ, at 35°C and a dilution rate of 0.21 h^{-1}. Complementing strain evaluation, the effects of cellular mass and sugar concentration were studied, taking into account the fact that these factors influence productivity calculations. In Figure 1 the kinetics of alcohol production can be compared and it can be seen that maximum alcohol production is reached at about 5 hr, except for the treatments with 30% sugar and 20 and 30% yeasts (66 and 99 g/ℓ dry cell weight, respectively), which were lower.

The complexity of yeast characteristics for alcohol production required the development of breeding methods other than hybridization in order to more efficiently achieve desired genetic gains. The adaptation of plant breeding methods to microbial strain improvement has been studied by Tavares.[22] Yeasts are amenable to such breeding schemes for several reasons: (1) they are amenable to selection in any ploidy level, (2) they have discrete mating among single cells which can be analyzed, (3) they have the advantage of tetrad analysis, (4) yeasts have a high recombination level, and (5) large numbers of individuals or popu-

Table 6
FERMENTATION KINETICS OF SELECTED HYBRIDS AND THE SEGREGANT M 304-2C (ETHANOL VALUES ARE IN °GL)

	Hours							
Strains	**1**	**2**	**3**	**4**	**5**	**6**	**7**	**8**
M 300 A	1.57	3.32	5.98	7.43	8.67	9.97	10.09	10.35
M 304-2C	1.83	3.74	5.51	7.64	9.23	10.23	10.33	10.88
H O 65	1.54	3.18	4.84	7.07	8.66	9.61	10.34	10.89
H O 82	1.29	2.92	4.08	5.81	7.74	8.77	9.58	9.66
H O 90	1.24	2.46	3.41	5.64	6.93	7.91	8.76	9.28
H 101	1.51	3.00	4.40	6.50	7.89	8.84	9.47	9.67
H 122	1.65	3.07	4.54	6.31	7.78	9.18	9.81	10.01

	1	**2**	**3**	**4**	**5**	**6**	**7**	**8**	**9**
M 300 A	2.57	3.09	5.08	6.10	7.20	9.08	9.90	11.05	11.05
M 304-2C	2.24	3.90	6.34	7.86	9.51	10.65	11.65	12.35	12.47
H O 65	2.20	3.01	5.15	6.22	7.93	9.78	10.21	11.22	12.10
H O 82	1.81	2.68	5.30	6.79	8.44	9.68	10.28	11.07	11.36
H O 90	2.34	3.14	5.24	6.10	7.88	9.01	10.07	10.97	11.14
H 101	2.21	2.86	4.89	5.82	7.32	8.59	9.92	10.61	11.19
H 122	2.45	3.25	5.16	6.45	8.05	9.42	10.63	11.47	11.57

	1	**2**	**4**	**5**	**6**	**7**	**8**
M 300 A	1.77	2.63	5.30	6.74	7.41	7.83	8.61
M 304-2C	2.15	3.25	5.97	7.62	8.22	8.55	9.27
H O 65	1.99	3.01	6.35	7.44	8.26	8.85	9.36
H O 82	2.16	3.24	6.45	7.44	8.10	8.36	8.74
H O 90	1.96	3.03	5.99	7.24	7.78	8.43	9.05
H 101	2.06	3.43	6.57	7.48	8.28	8.65	9.44
H 122	2.08	3.19	6.39	7.09	8.29	8.65	9.54

	Alcohol concentration after 12 hr fermentation (°GL)			
Strains	**I**	**II**	$\overline{X}$	**g/100 mℓ**
M 300 A	14.66	12.92	13.79	10.88
M 304-2C	14.04	14.05	15.045	11.09
H O 65	14.12	13.47	13.795	10.89
H O 82	13.94	13.34	13.64	10.77
H O 90	14.20	13.70	13.95	11.01
H 101	13.35	13.00	13.17	10.40
H 122	13.77	13.26	13.515	10.68

lations can be handled with ease. In summary, the genetic basis of quantitative characteristics can be properly analyzed better than any other fungal system.[12]

A technique was developed to handle large numbers of strains and cells similar to the polycross test and other methods used in plant breeding.[23] Figure 2 shows the adaptation of such a scheme to heterothallic and homothallic yeasts. The first step in the breeding program is to perform a mass mating among strains followed by sporulation, followed by heat treatment to kill nonsporing cells and another mass mating. The cells are then plated and larger colonies are selected for further evaluation by growth under high temperature and on different sugar sources. From about 50 colonies, 20% are generally selected, and a second

Table 7
KINETIC PARAMETERS FOR SELECTED HYBRIDS IN DISCONTINUOUS FERMENTATION WITH CELL RECYCLE

Strains	X(g/ℓ)	S(g/ℓ)	Efficiency (g/g)	Productivity P(g/ℓ hr)	Productivity up(g/g · hr)	Etanol max 12 hr (g/ℓ)
M 300 A	30.3	232.0	0.47	10.93	0.36	108.8
M 304-2C	30.1	232.0	0.49	14.67	0.49	110.9
H O 65	30.3	232.0	0.47	13.03	0.43	108.9
H O 82	30.3	232.0	0.46	15.20	0.50	107.7
H O 90	30.5	232.0	0.47	12.43	0.41	110.1
H 101	30.3	232.0	0.45	11.70	0.39	104.0
H 122	29.8	232.0	0.46	12.67	0.42	106.8

Table 8
THE EFFECTS OF TEMPERATURE AND OSMOTIC PRESSURE (15—25% SUGAR) ON ETHANOL PRODUCTION (E) AND YIELDS (R) OF SELECTED HYBRIDS (ETHANOL VALUES IN °GL)

	Temp and reducing sugars							
	35°C — 20%		35°C — 20%		40°C — 20%		40°C — 25%	
Strains	E	R	E	R	E	R	E	R
M 300 A	10.2	(80)	10.2	(64)	6.1	(48)	5.0	(31)
FF	8.8	(76)	8.6	(54)	4.6	(36)	4.3	(27)
H O 65	9.5	(74)	10.3	(64)	6.2	(49)	5.3	(33)
H O 73	10.1	(79)	10.6	(66)	5.3	(41)	4.4	(28)
H O 81	9.3	(73)	10.7	(67)	5.6	(44)	5.3	(33)
H O 82	10.1	(79)	11.3	(71)	5.7	(45)	5.4	(34)
H 101	10.9	(85)	10.6	(66)	6.5	(51)	6.5	(41)
	35°C — 15%		35°C — 20%		40°C — 15%		40°C — 20%	
	E	R	E	R	E	R	E	R
M 300 A	6.6	(65)	7.6	(59)	5.5	(54)	2.0	(15)
FF	6.9	(68)	7.8	(60)	4.9	(48)	2.8	(22)
H O 13	7.0	(69)	7.6	(59)	4.3	(42)	3.0	(23)
H O 48	6.9	(68)	8.4	(65)	4.4	(43)	5.1	(39)
H O 58	7.6	(74)	8.5	(66)	6.2	(61)	2.6	(20)
H O 79	6.1	(60)	7.0	(54)	5.1	(50)	4.6	(36)
	35°C — 20%		35°C — 25%		40°C — 20%		40°C — 25%	
	E	R	E	R	E	R	E	R
M 300 A	9.9	(71)	8.6	(51)	5.5	(39)	3.4	(20)
FF	8.8	(63)	7.3	(43)	4.4	(31)	2.4	(14)
H O 84	9.2	(66)	7.6	(45)	5.0	(36)	3.0	(18)
H O 85	10.3	(73)	7.3	(43)	3.6	(26)	2.9	(17)
H O 90	11.1	(79)	8.0	(47)	4.4	(31)	4.3	(25)
H 106	8.8	(63)	6.6	(39)	4.5	(32)	3.7	(22)
H 122	10.1	(72)	8.5	(50)	5.9	(42)	6.0	(35)

Table 9
EVALUATION OF M 304-2C STRAIN IN DISCONTINUOUS FERMENTATION

Batches	Hours	Ethanol °GL	Ethanol g/100 mℓ	Yields (%)
I	15	6.4	5.10	47.4 (aeration)
II	6	12.6	9.99	93.3
III	15	13.1	10.37	97.0
IV	7	12.5	9.87	92.6
V	24	12.0	9.49	89.0
VI	20	12.4	9.76	91.8
VII	15	7.0	5.58	52.2 (aeration)
VIII	8	10.4	8.22	77.3
IX	15	11.4	9.02	84.4
X	8.5	10.9	8.62	80.7
XI	15	10.9	8.62	80.7
XII	8.5	12.3	9.69	91.0
XIII	15	11.3	8.89	90.9

qualitative evaluation is made selecting those with better fermentation than the parental strains. Upon final selection, they are sporulated and mass mated again to start another selection round. An alternative route to this method is to avoid plating, and to perform a fermentation after mass mating for two or more times under specific stress conditions, then sporulate the cells, heat treat and mass mate the cells, followed by further fermentations. This alternative route can be repeated various times as in the bulk selection method of plant breeding.[23]

Figure 3 and Table 10 show the results of the polycross method after three rounds. The joint evaluation of the strains selected in every polycross round has shown that 20% production above industrial tester strains is achieved. The best strains selected after joint evaluation were tested again in the presence of four tester strains, as the results presented in Table 11 show. The best strain is 39% better than the standard Fleischmann yeast and it is clear that high producing strains and tolerance to stressful conditions can be obtained after only three polycross rounds. For instance, mean values (g/100 mℓ) for test strains, first, second, and third rounds of selection were 8.218, 8,296, 8.834*, and 9.404**, respectively (Table 11). Despite the elevated variation coefficient (11.7%) obtained for this experiment, significant increases were detected by Dunnett's test comparing polycross round with grouped tester means. Relating means and rounds by regression analyses, a significant linear mean gain of 6.68% was obtained per round in relation to the original mean.

Another breeding method that has been adapted to yeast is selection after self-crossing to choose superior recombinant haploid strains and hybrids. The proposed scheme presented in Figure 4 starts from a superior hybrid for fermentation and one which is able to sporulate as required for the polycross breeding method. After heat treatment and plating of cells about 50 (10%) larger colonies are visually selected. They are evaluated in fermentation tests and the best 20% selected to be sporulated and mass mated, thus starting another round. This method was applied starting from two different hybrids and stopping after five self-crossing rounds. One of the hybrids (E 121) of triploid genetic make-up nature presented a mean gain of 16.1% and the diploid strain (E 202) a mean gain of 6.4%. This difference in gain is due to the fermentation potential of the hybrids. The diploid fermented simply by chance, better than the triploid strain, and selection was not very efficient in every round. The estimated variability and heritability values decreased with the number of selection rounds, while the heritabilities for the triploid strain were about half the value for the diploid.

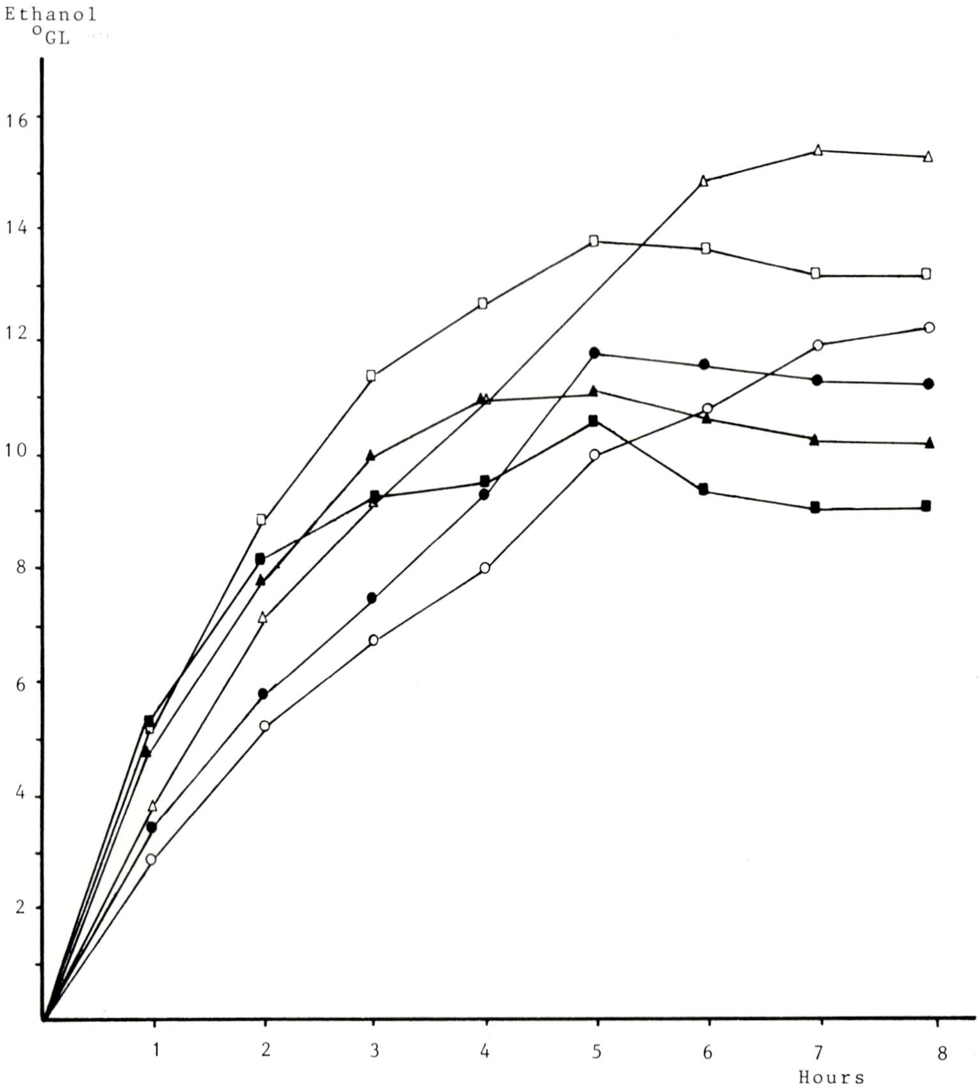

FIGURE 1. The effects of sugar concentration (S_o) and dry cell weight (X_o) on fermentation kenetics of M 304-2C strain. —▲— S_o = 200 g/ℓ, X_o = 99 g/ℓ; —△— S_o = 300 g/ℓ, X_o = 99 g/ℓ; —■— S_o = 200 g/ℓ, X_o = 132 g/ℓ; —□— S_o = 300 g/ℓ, X_o = 132 g/ℓ; —●— S_o = 200 g/ℓ, X_o = 66 g/ℓ; —○— S_o = 300 g/ℓ, X_o = 66 g/ℓ.

From five rounds heritability, a mean value of 0.448 was obtained from strain E 121 (triploid) and a mean value of 0.781 was obtained for strain E 202 (diploid).

Inbreeding is one way to obtain strains of decreased genetic variation. Another way is to perform back-cross breeding or recurrent selection[23] techniques. They are of value in obtaining isogenic strains of different mating types, adequate strains for special experiments like transformation, and as a useful technique for the purpose of conducting quantitative genetic studies. As seen in Figure 5, from two haploid strains M 304-2C and X 2180-1B a hybrid was obtained, sporulated, and single spores separated from individual tetrads by micromanipulation. Mating-type scoring was performed and a spore bearing the α mating type taken randomly was crossed with the recurrent parent M 304-2C. The new hybrid was then sporulated and a second back-cross round started. After four back crosses it was clear

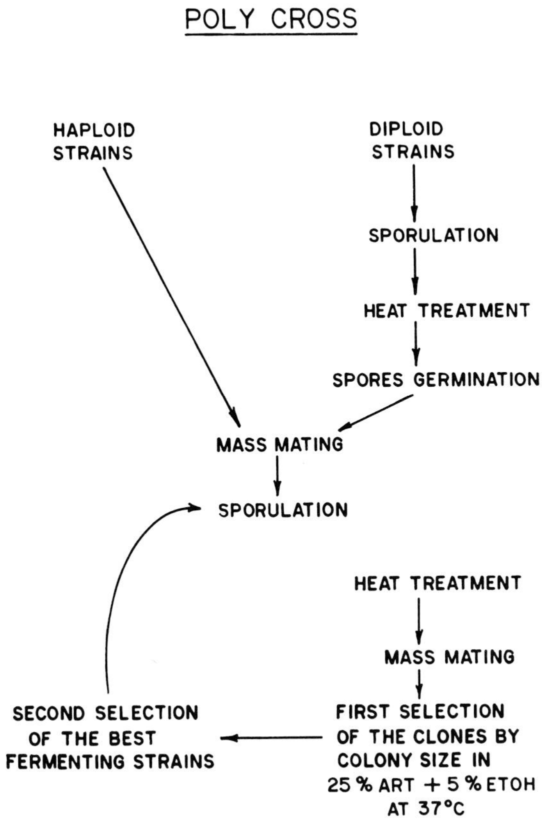

FIGURE 2. The polycross scheme adapted to homothallic and heterothallic yeasts for the breeding of quantitative characters.

that the tendency after every cross was in the direction of the recurrent parent, as demonstrated by the mean values of alcohol production obtained (Figure 6). In Table 12, ethanol production values achieved for every round of back crossing can be seen. As seen in Table 13, the segregation of MX1 hybrid has shown the highest degree of variation, while ethanol production for sporulating strains is variable with an F test value of 2.83*. The first and second back crosses show that a high degree of uniformity and variability is lost (F test values, respectively, of 1.72 and 1.01 nsec). From comparison of mean values, no significant differences can be seen between ascus means and general segregant means, which suggests additive variation for the character.

The breeding methods were developed with the goal of exploiting natural variability, and care must be taken not to exclude the possibility of utilizing mutation-selection and other techniques, especially when haploids are involved. These methods are being analyzed in order to estimate specific quantitative genetic parameters, with the objective of understanding the characteristics of productivity and production yield potential in alcohol fermentation.

IV. ETHANOL PRODUCTION FROM POLYFRUCTOSANS OF *HELIANTHUS TUBEROSUS* L

High concentrations of inulin and other polyfructosans in *Helianthus tuberosus* require yeasts with inulinase activity and high product yields for the industrial processing of this raw material to obtain ethanol and fructose concentrates. Many studies have been conducted on this subject.[24-28] We selected several strains from over 21 species[29] with the aim of

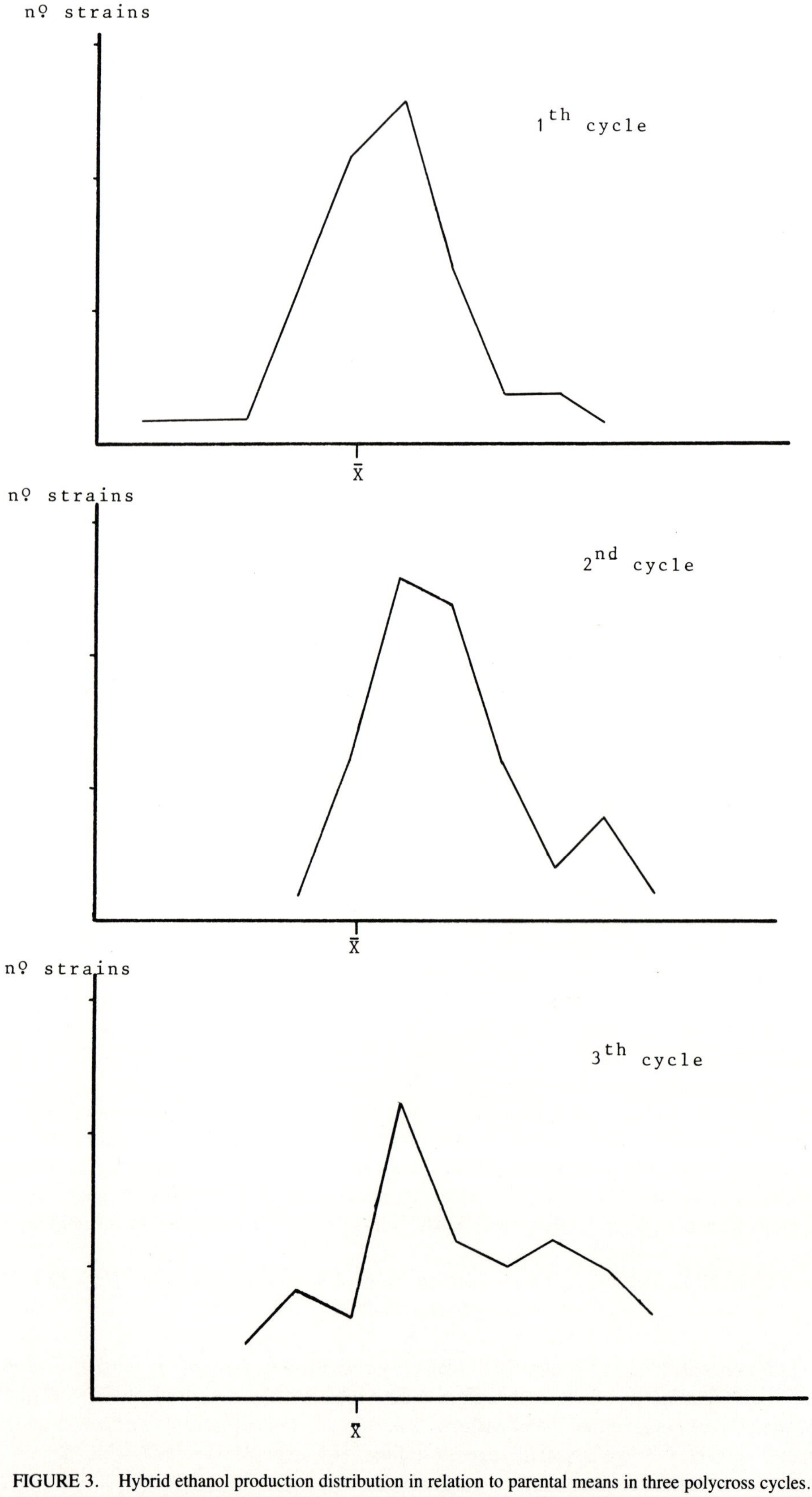

FIGURE 3. Hybrid ethanol production distribution in relation to parental means in three polycross cycles.

Table 10
JOINT EVALUATION OF SELECTED STRAINS FROM THREE ROUNDS OF POLYCROSS SELECTION

Tester and parental strains		First	round	Second	round	Third	round
M 300 A	8.083	P-I-3	8.175	P-II-1	9.791	P-III-2	9.424
Fleishmann	8.715	P-I-4	7.788	P-II-5	8.834	P-III-4	9.756
E 202-IV-29	9.357	P-I-5	8.034	P-II-23	9.041	P-III-5	8.278
Montrachet	7.963	P-I-15	8.672	P-II-30	7.596	P-III-6	10.445
α	9.073	P-I-18	9.068	P-II-31	8.754	P-III-9	7.966
J 132 b	7.972	P-I-19	8.378	P-II-34	8.517	P-III-11	8.256
FI-A	7.795	P-I-31	8.924	P-II-35	8.404	P-III-12	9.160
F1-C	8.184	P-I-37	9.048	P-II-36	9.498	P-III-14	9.967
IZ 671	7.460	P-I-41	7.919	P-II-40	9.295	P-III-17	10.509
IZ 672	6.919	P-I-43	6.950	P-II-45	8.705	P-III-18	10.279
ATCC 4126	7.482						
ATCC 4132	9.151						
ATCC 26603	8.877						
ATCC 24858	8.015						

Note: Ethanol values in °GL.

Table 11
JOINT EVALUATION OF THE SUPERIOR YEAST STRAINS FROM THREE ROUNDS OF SELECTION USING THE POLYCROSS METHOD

	Fermentations							
Strains	I	II	III	IV	V	$\overline{X}$	SD	Yield
M 300 A	7.95	6.61	6.53	6.68	6.65	6.88	±1.20	70.6 ± 12.3
Fleishmann	6.61	5.67	5.74	6.23	6.49	6.15	±0.85	63.1 ± 8.7
E 202-IV-39	8.66	5.56	6.94	7.34	7.29	7.16	±2.22	73.5 ± 22.8
M 304-2C	8.22	8.19	7.36	7.02	7.05	7.57	±1.19	77.7 ± 12.2
P-I-37	8.03	6.95	7.14	7.72	7.60	7.49	±0.88	76.9 ± 9.0
P-II-1	8.50	6.81	7.02	7.86	7.61	7.56	±1.35	77.6 ± 13.8
P-II-5	8.22	8.34	7.99	7.99	8.07	8.12	±0.31	8.3. ± 3.1
P-II-36	8.58	4.89	5.86	6.69	6.39	6.48	±2.72	66.5 ± 27.8
P-III-4	8.62	7.41	7.48	7.80	7.91	7.84	±0.96	80.5 ± 9.8
P-III-6	8.50	7.29	7.21	7.72	7.52	7.65	±1.03	78.5 ± 10.5
P-III-12	8.58	7.25	8.03	8.35	8.43	8.13	±1.06	83.4 ± 10.9
P-III-14	8.54	7.44	7.41	8.42	7.57	7.88	±1.11	80.9 ± 11.4
P-III-17	8.46	7.18	7.17	7.64	7.80	7.65	±1.06	78.5 ± 10.9
P-III-18	8.62	8.34	8.22	8.90	8.74	8.56	±0.56	87.9 ± 5.7

Note: Values are presented in g/100 mℓ.

developing yeasts for polyfructosans fermentation. A strain of *Kluyveromyces marxianus* after successive fermentation reached 85% of product yield in a nonagitated fermentation conducted at 30°C with 15.6% of total reducing sugars and a 48-hr fermentation time. After 10 hr, under such conditions, yields of 75.4% were obtained with a volumetric rate of 5.9 $g/\ell \cdot h^{-1}$. In stirred fermentation, alcohol production was significantly increased with a concentration of 78 g/ℓ (X), reaching a productivity of 11.17 $g/\ell \cdot hr$, with a product yield of 96.83% after 13 hr. Figure 7 presents comparative values of alcohol production and sugar attenuation of extracts from Jerusalem artichokes by *K. marxianus* and Fleischmann yeasts.

Besides breeding yeast for alcohol production, we are interested in the genetics of *Kluy-*

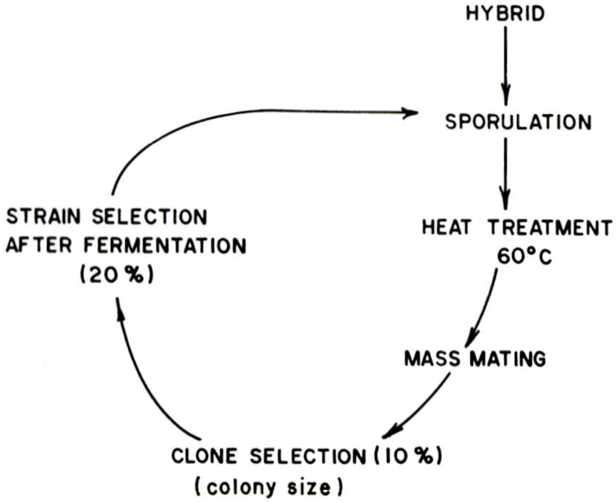

FIGURE 4. Selection after self-crossings (inbreeding) scheme adapted to yeast breeding for polygenic characters.

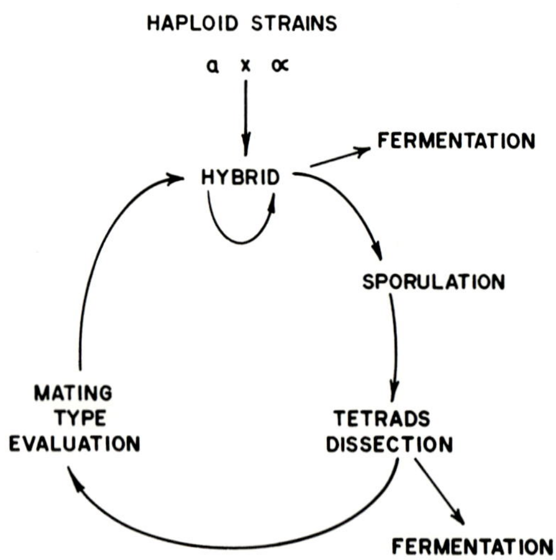

FIGURE 5. Back-cross breeding scheme adapted to yeast improvement.

veromyces yeasts. An understanding of genetics will help to better assess enzyme production and other important characteristics for fermentation. In this vein, auxotrophic mutants were obtained from different strains of *K. marxianus, K. fragilis,* and *K. lactis,* and hybrids were obtained after performing natural crosses by rare mating and protoplast fusions when necessary. Though good hybrid stability was found in intra- and interspecific crosses, alcohol production was not affected significantly. *K. marxianus* × *K. lactis* hybrids, obtained by protoplast fusion, were shown to be less productive than the respective parental strain for lactose fermentation. This was also true for fermentation of polyfructosans; though some of them were better in fermenting glucose (Table 14). The alcohol production decrease observed for some hybrids in fermentation of certain carbon sources was also seen for other amylolytic strains.

Segregation analyses of *K. marxianus* crosses are under way and some markers are showing linkage.[39] We have quantitated analysis of variation among segregants and have seen a level of significance which would indicate the chances for breeding for characteristic for the production of ethanol from fructosans to be 53.37%.

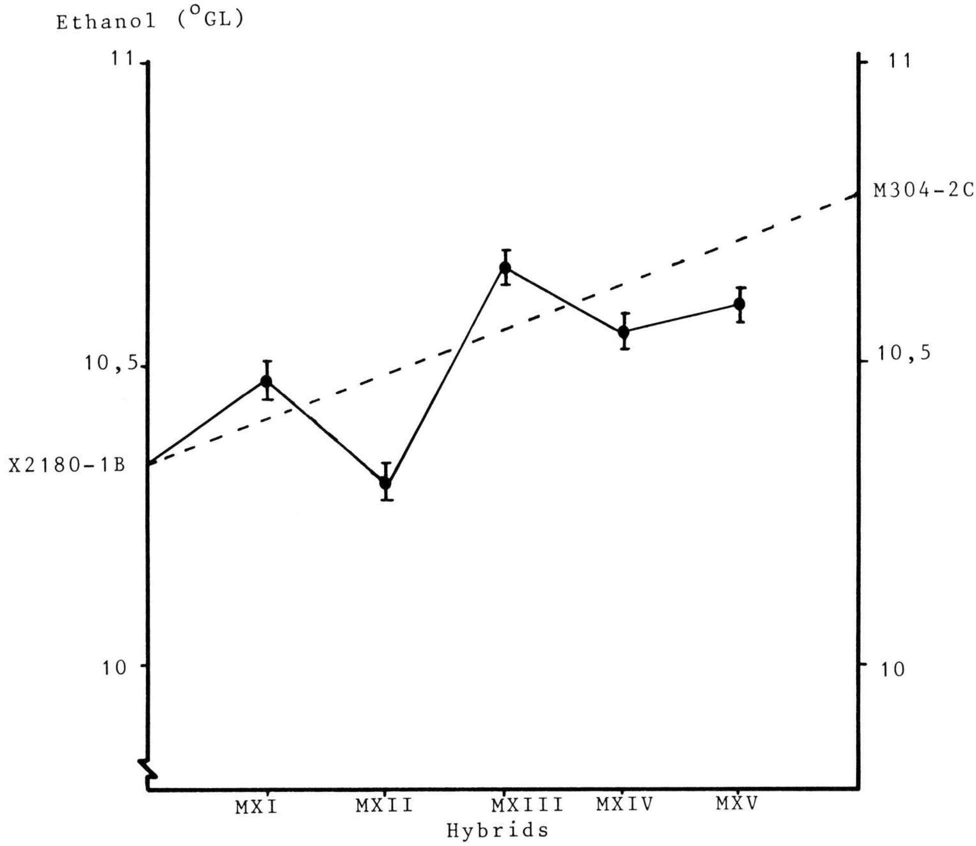

FIGURE 6. Ethanol production by hybrids obtained from back-cross breeding involving four back-cross cycles.

Table 12
ETHANOL PRODUCTION IN BACK-CROSS BREEDING

Strains	Ethanol (°GL)	Ethanol productivity ($g/\ell \cdot h^{-1}$)	Mean variation for successive fermentations (%)
P_1 (M 304-2C)	10.79 ± 0.09	5.96	0.80
P_2 (X 2180-1B)	10.34 ± 0.43	5.46	4.69
M × I	10.48 ± 0.34	5.50	3.25
M × II	10.31 ± 0.33	5.15	3.24
M × III	10.65 ± 0.27	5.44	2.51
M × IV	10.55 ± 0.22	5.72	2.12
M × V	10.60 ± 0.19	5.55	1.85

Note: Parental, hybrid, and back crosses mean ethanol production is given.

Table 13
ALCOHOL PRODUCTION (°GL) FOR PARENTAL, HYBRID, AND F_2 STRAINS

				Tetrads									
Parental		Hybrid		1		2		3		4		5	
X2180-1B	9.07			A	10.84	A	9.22	A	9.09	A	9.41	A	9.69
		MX 1	10.12	B	7.72	B	8.69	B	9.62	B	9.50	B	9.16
M304-2C	10.37			C	8.77	C	9.42	C	9.68	C	9.30	C	8.50
				D	10.34	D	10.08	D	8.27	D	10.08	D	9.99

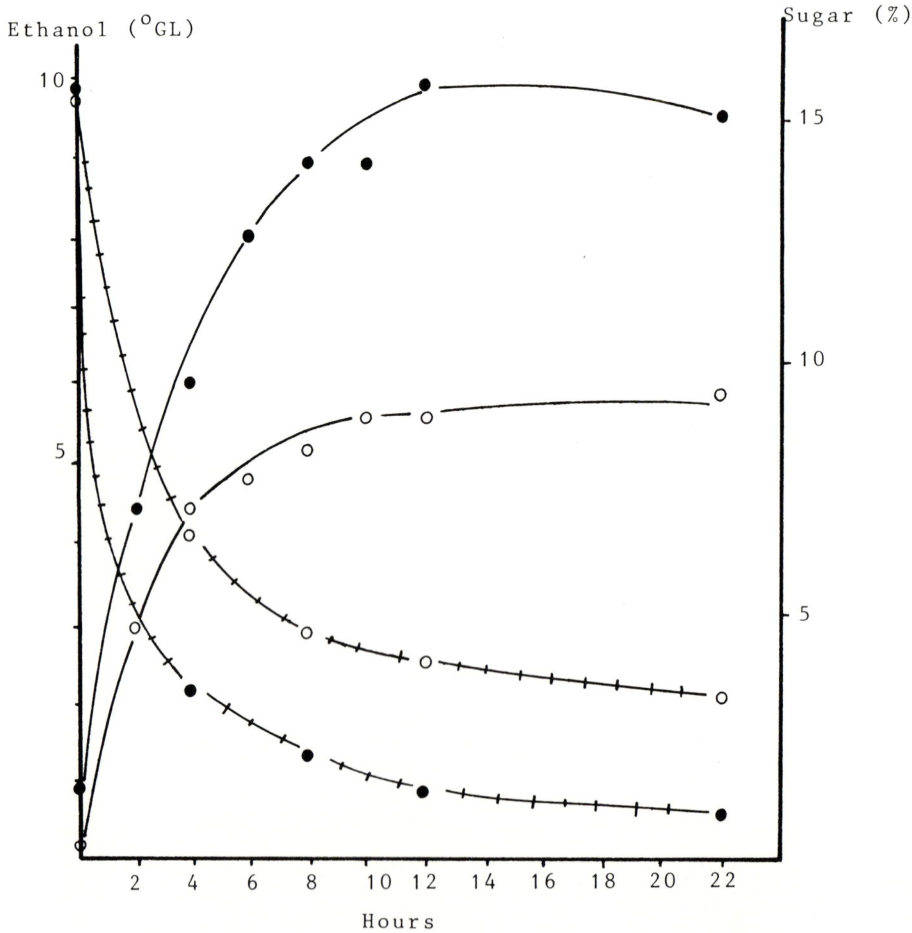

FIGURE 7. Ethanol production and sugar attenuation from extracts of Jerusalem artichoke by *Kluyveromyces marxianus* (○—○) and *Saccharomyces cerevisiae* (●—●).

For the strains analyzed previously the maximum inulinase activity obtained was 3.6 μ*M* (fructose equivalent)/mℓ/min, while catabolic repression of inulinase by fructose was detected in both *Kluyveromyces* yeasts,[29] however, a detailed analysis has revealed that catabolic repression by glucose was pronounced in *K. fragilis* and that enzyme induction by inulin was observed. In *K. marxianus* constitutive enzyme production was observed and a stimulating effect by inulin for the enzyme production was seen (Table 15).

Table 14
EVALUATION OF FERMENTATION CAPABILITIES OF FUSION PRODUCTS FROM *KLUYVEROMYCES MARXIANUS* × *K. LACTIS* FUSIONS

Strains	13% Glucose	13% Jerusalem artichoke extract	3% Lactose
Fleischmann	7.734 ± 0.152	3.963 ± 0.315	0.000
IZ 1339	6.683 ± 0.201	5.796 ± 0.393	0.000
KM-9	5.683 ± 0.237	5.902 ± 0.393	0.000
ATCC 8585	6.171 ± 0.111	2.283 ± 0.112	1.340 ± 0.038
FML-04	5.064 ± 0.204	5.743 ± 0.342	1.158 ± 0.121
FML-05	6.780 ± 0.103	5.356 ± 0.204	0.706 ± 0.209
FML-09	7.311 ± 0.099	5.233 ± 0.094	1.026 ± 0.264
FML-11	6.641 ± 0.215	5.530 ± 0.209	0.258 ± 0.125

Note: Alcohol values are in °GL, being the mean of three consecutive fermentations.

Table 15
INULINASE ACTIVITY OF *KLUYVEROMYCES MARXIANUS* AND *K. FRAGILIS* STRAINS WITH DIFFERENT CARBON SOURCES

	Strains			
	K. marxianus		*K. fragilis*	
Fermentation media	IZ 1339	IZ 619	IZ 610	IZ 275
Inulin (0.5%)	458.4	1815.5	400.3	305.3
Sucrose (0.5%)	449.3	1399.4	117.26	90.98
Glucose (0.5%)	367.6	960.7	54.58	0.00
Celobiose (0.5%)	621.8	1647.6	18.20	11.00
Ethanol (0.5%)	351.7	771.6	18.20	0.00
Inulin (0.5%) + glucose (0.5%)	526.5	1929.0	113.22	200.16
Sugarless	365.0	1191.4	38.40	0.00

Note: Values expressed in units (n · mol fructose/mℓ culture medium · min^{-1})

V. BREEDING FOR STARCH CONVERSION TO ETHANOL

Although sucrose from sugar cane will predominate as the carbon source in the alcohol industries in Brazil, much academic research has been conducted in starch conversion to ethanol; indeed, new plants are presently in operation using this process. Amylolytic yeasts and local production of amylolytic enzymes are not used, although some microorganisms favorable to commercial production have been obtained from previous work.[19]

In order to develop yeasts for starch fermentation, a program was initiated with the objective of evaluating different selection techniques for hybrid recovery from natural crosses and protoplast fusion, in addition to analyzing the fermentation ability and genetic stability of interspecific hybrids. The initial studies involved multiple interspecies crosses and hybrids

Table 16
FERMENTATION ABILITY OF *SACCHAROMYCES DIASTATICUS* × *S. CEREVISIAE* HYBRIDS WITH DIFFERENT SUBSTRATES

Fusion hybrids	Dextrinized starch	Sugar cane molasses	Natural crosses	Dextrinized starch	Sugar cane molasses
P_1 (57-5)	4.45	7.76	P_1 (J132b)	5.43	6.59
P_2 (C-74)	1.00	2.76	P_2 (F1-A)	1.65	6.66
H_1 (CD 301)	4.15	7.74	H_1 (E 201)	3.21	6.88
H_2 (CD 336)	4.35	7.43	H_2 (E 202)	2.58	6.92
H_3 (CD 342)	4.41	7.85	H_3 (E 203)	2.08	6.75
H_4 (CD 487)	4.13	7.16			
H_5 (CD 682)	3.68	7.93			
H_6 (CD 840)	2.43	7.69			

Note: Alcohol values in g/100 mℓ.

Table 17
FERMENTATION OF POLYPLOID HYBRIDS FROM *SACCHAROMYCES DIASTATICUS* × *S. UVARUM* IN TWO DIFFERENT SUBSTRATES

Strains	Ploidy, genotype	Dextrinized starch	Sugar cane molasses
P_1 (J 132 b)	n a DEX1	3.09	6.73
H_1 (E 104)	2n a/αDEX1 dex	2.03	7.34
H_2 (E 106)	3n a/ααDEX1 dexdex	1.37	7.51
H_3 (E 107)	5n a/ααααDEX1 dexdexdexdex	1.11	6.70
H_4 (E 120)	3n a/aα DEX1 DEX1 dex	2.29	7.20
H_5 (E 121)	3n a/aα DEX1 DEX1 dex	2.15	7.32
H_6 (E 140)	4n a/aαα DEX1 DEX1 dex dex	1.23	6.77
H_7 (E 141)	4n a/aαα DEX1 DEX1 dex dex	1.22	6.95
H_8 (E 150)	6n a/aαααα DEX1 DEX1 dexdexdexdex	1.93	7.23
H_9 (E 151)	6n a/aαααα DEX1 DEX1 dexdexdexdex	2.08	7.10

Note: Alcohol values in g/100 mℓ.

were obtained from crosses between *Saccharomyces diastaticus, S. cerevisiae, S. uvarum,* and *S. baillii.* These hybrids exhibited starch fermentation abilities which were equal to or lower than the parental strain, *S. diastaticus.* The results were independent of the selection technique or the parental strain.[30] On the other hand, the hybrids did present positive heterosis for sugar cane molasses fermentation when parental strains were themselves good fermenters (Table 16). Thus, the genetic diversity between parental strains can be used for breeding purposes, as suggested by Lindegren.[10]

Polyploid hybrids were obtained from *S. diastaticus* and *S. uvarum.* Negative heterosis was observed for dextrinized starch fermentation with these hybrids, though they presented positive heterosis in sugar cane molasses fermentation (Table 17). For triploids it was easy to compare the effect of gene dosage, as seen by Stewart.[31] From *S. baillii* × *S. diastaticus* three hybrids were selected and they presented heterosis of intermediate value compared to parental values in fermentating starch. Similar results have been previously observed.

Many yeasts were analyzed for α-amylase and glucoamylase activity and a strain of *Schwanniomyces castellii* was the best, with 398 and 28.34 U/mℓ of α-amylase and glucoamylase activity, respectively, after 36 hr of growth. Derepressed mutants[32,33] were obtained for *S. castellii, S. alluvius,* and *Saccharomyces diastaticus* after selection for resistance

Table 18
EVALUATION OF DEREPRESSED MUTANTS FOR STARCH CONVERSION

Schwanniomyces castelli		*S. alluvius*		*Saccharomyces diastaticus*	
Strains	**Conversion (%)**	**Strains**	**Conversion (%)**	**Strains**	**Conversion (%)**
Original (58-7)	0.00	Original (54-103)	0.00	Original (5132 b)	0.00
M_1	36.51	M_1	31.30	M_1	19.50
M_2	96.82	M_2	0.00	M_2	0.00
M_3	96.82	M_3	55.50	M_3	0.00
M_4	0.00	M_4	50.70		
M_5	98.41	M_5	95.60		
M_6	80.16	M_6	34.30		
M_7	19.05	M_7	52.30		

Note: Values expressed as percent starch hydrolyzed in the presence of glucose (0.5% starch + 2.5% glucose).

to 2-deoxy D-glucose. Resistance to this drug was not necessarily followed by derepression and among the derepressed strains different derepression levels were observed (Table 18). As can be observed in Figure 8, immediate conversion of starch even occurred in the presence of glucose.

Considering the properties of *Schwanniomyces castellii,* this yeast was chosen as a source of amylolytic enzyme genes for cloning into good fermentation yeast strains. A genomic library was prepared from a derepressed strain of *S. castellii* in *Escherichia coli* and transformed cells are currently being evaluated for gene expression.

VI. THE USE OF SINGLE GENETIC FACTORS

Among major genes or cytoplasmic factors there are basically two alternatives of interest to industry. One is to have major genes for the purpose of identifying strains without compromising industrial performance, and the other is to work with such markers to improve specific physiological properties. Killer factor[34-37] and flocculence[31] are characteristics of particular interest, since it is possible to take advantage of the killing property in order to eliminate sensitive contaminants from natural flora and to use the natural settlement of cells in alternative fermentation processes without utilizing equipment for cell recovery. Flocculence is particularly interesting for small-scale plants where yeast cultures are recycled. Many yeasts from various species and genera have been screened for the presence of the killer characteristic, and the most effective were found in *Saccharomyces diastaticus, S. cerevisiae,* and *Kluyveromyces marxianus.* Cross tests for sensitivity to the killer toxin were evaluated among these killer strains. The *K. marxianus* killer was found to eliminate *S. cerevisiae* and *S. diastaticus* killer strains and the *S. diastaticus* had a killer factor different from that of *S. cerevisiae.* Due to the absence of killer tester strains from *S. cerevisiae,* further work will be necessary to distinguish the killer plasmids in each strain.

Because the killer from *K. marxianus* presented the most effective killing effect over many yeasts from different species, its transfer has been made to other *Kluyveromyces* species. In *K. lactis,* full expression of the killer has been obtained, though with *K. fragilis* hybrids did not initially express the marker. It reappeared, however, when segregants from these hybrids were obtained and showed full expression. In some ways toxin production is inhibited by transformant *K. fragilis* with no plasmid loss. This phenomenon is under evaluation.

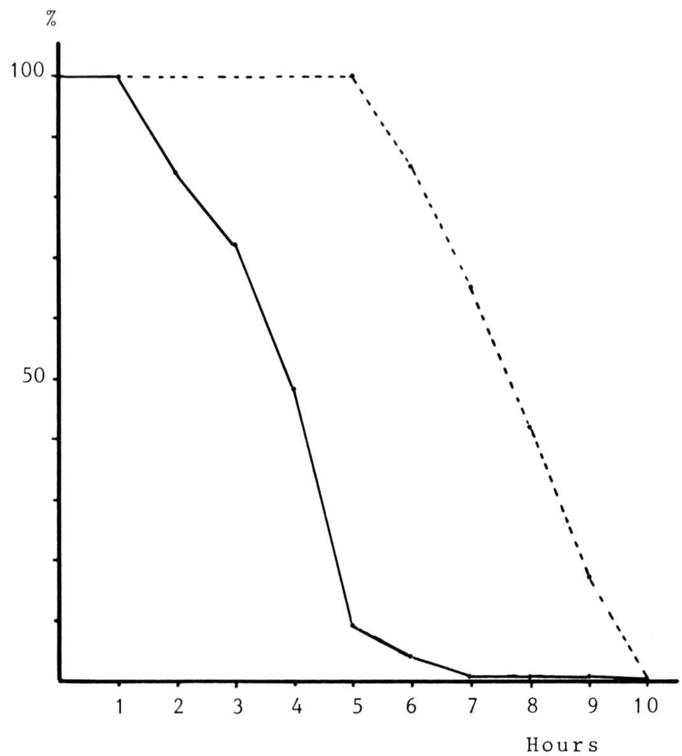

FIGURE 8. Starch conversion by normal (58-5) and derepressed strains (58-5-11-5) of *Schwanniomyces castellii* in media with 2.5% glucose plus 0.5% starch. *S. castellii:* ---------58-5 (original strains); ——58-5-11-5 (derepressed mutant).

Little success was obtained with *Saccharomyces* × *Kluyveromyces* hybrids due to high mitotic instability. When *Kluyveromyces* cells segregate, expression of the killer factor is seen. However, expression of killer phenotype has not been observed when *Saccharomyces* types of cells segregate, which suggests occurrence of plasmid loss during mitotic segregation of the hybrid.

Flocculence transferred to some yeast has provided the opportunity to develop a tower-type fermentation process.[38] One of the hybrids obtained for this purpose in continuous fermentation yielded productivity of 28.6 g/ℓ h^{-1} and 10 g/100 mℓ alcohol production at 35°C. Tests are running on an industrial pilot plant scale where good agreement with small unit values has been obtained.

In summary, it is quite evident from the data presented above that for constructing yeast strains for usage in industry, a traditional approach involving properly planned breeding experiments can be of great value. In spite of the recent advances in recombinant DNA technology, techniques such as Hybridization Breeding, Recurrent Selection, Polycross Breeding, and Selection after Self-Crossings present a viable and effective means of constructing industrial yeast strains possessing the desired characteristics.

ACKNOWLEDGMENTS

The authors wish to express gratitude to Ana Brancalion, Luiz Gomes, and Paulo Ortolan for their assistance during the course of the research described in this study. Acknowledgments are also due to Fundo de Incentivo à Pesquisa Técnico-Cientifica (FIPEC-BB), Financiadora de Estudos e Projetos (FINEP-SEPLAN), and Fundação de Amparo à Pesquisa no Estado de São Paulo (FAPESP) for the financial support, and to Usina Modelo Açucar e Alcool S.A. where industrial parameters were evaluated.

REFERENCES

1. **Nagodawithana, T. W., Castellano, C., and Steinkraus, K. H.,** Effect of dissolved oxygen, temperature, initial cell count and sugar concentration on the viability of *Saccharomyces cerevisiae* in rapid fermentations, *Appl. Microbiol.*, 28, 383, 1974.
2. **Thomas, D. S., Hossach, J. A., and Rose, A. H.,** Plasma membrane lipid composition and ethanol tolerance in *Saccharomyces cerevisiae, Arch. Microbiol.*, 117, 239, 1978.
3. **Ho, N. W. Y.,** Yeast alcohol tolerance and recombinant DNA for improved alcohol production process, in *Annual Reports on Fermentation Processes*, Vol. 4, Tsao, G. T., Ed., Academic Press, New York, 1980, 235.
4. **Panchal, C. J. and Stewart, G. G.,** The effect of osmotic pressure on the production and excretion of ethanol and glycerol by a brewing yeast strain, *J. Inst. Brew.*, 86, 207, 1980.
5. **Haraldson, A. and Bjorling, T.,** Yeast strains for concentrated substrates, *Eur. J. Appl. Microbiol. Biotechnol.*, 13, 34, 1981.
6. **Novak, M., Strehaiano, P., Moreno, M., and Goma, G.,** Alcoholic fermentation: on the inhibitory effect of ethanol, *Biotechnol. Bioeng.*, 23, 201, 1981.
7. **Beaven, M. J., Charpentier, C., and Rose, A. M.,** Production and tolerance of ethanol in relation to phospholipid fatty-acyl composition of *Saccharomyces cerevisiae* NCYC 431, *J. Gen. Microbiol.*, 128, 1447, 1982.
8. **Hacking, A. J., Taylor, I. W. F., and Hanas, C. M.,** Selection of yeast able to produce ethanol from glucose at 40°C, *Appl. Microbiol. Biotechnol.*, 19, 361, 1984.
9. **Calam, C. T., Daglish, L. B., and Gaitskell, W. S.,** Hybridization experiments with *Penicillium patulum* and *Fusarium moniliforme*, in *Genetics of Industrial Microorganisms*, Vol. 2, Vanek, Z., Hostalek, Z., and Cudlin, J., Eds., Elsevier, Amsterdam, 1973.
10. **Lindegren, C. C.,** *The Yeast Cell, Its Genetic and Cytology*, Educational Publishers, St. Louis, 1949.
11. **Pateman, J. A. and Lee, T. O.,** Segregation of polygenes in ordered tetrads, *Heredity*, 15, 351, 1960.
12. **Jinks, J. L., Caten, C. E., Sinchen, G., and Croft, J. H.,** Heterokaryon incompatibility and variation in wild populations of *Aspergillus nidulans, Heredity*, 21, 227, 1966.
13. **Simchen, G.,** Monokaryotic variation and haploid selection in *Schizophyllum commune, Heredity*, 21, 241, 1966.
14. **Elander, R. P. and Chang, L. T.,** Microbial culture selection, in *Microbial Technology*, Vol. 2, Peppler, H. J. and Perlman, D., Eds., Academic Press, New York, 1979, chap. 12.
15. **Merrick, M. J. and Caten, C. E.,** The inheritance of Penicillin titre in wild-type isolates of *Aspergillus nidulans, J. Gen. Microbiol.*, 86, 283, 1975.
16. **Caten, C. E.,** Quantitative genetic variation in fungi, in *Quantitative Genetic Variation*, Thompson, J. N. and Thoday, J. M., Eds., Academic Press, New York, 1979, 35.
17. **Mather, K. and Jinks, J. L.,** *Biometrical Genetics. The Study of Continuous Variation*, 3rd ed., Chapman and Hall, London, 1982.
18. **Tavares, F. C. A.,** Yeast hybridization breeding for ethanol production, in Proc. 2nd Japan-Brazil Symp. on Science and Technology, 1980, 185.
19. **Tavares, F. C. A.,** Genetic engineering for ethanol production, in *Genetic Engineering for Biotechnology*, Crocomo, O., Tavares, F. C. A., and Sodzeieski, D., Eds., Grafica Editora Hamburg Ltda., São Paulo, Brazil, 1981.
20. **Wang, D. I. C., Cooney, C. L., Demanin, A. L., Dunnil, P., Humphrey, A. E., and Lilly, M. D.,** *Fermentation and Enzyme Technology*, John Wiley & Sons, New York, 1979.
21. **Maiorella, B., Wilke, C. R., and Blanch, H. W.,** Alcohol production and recovery, *Adv. Biochem. Eng.*, 20, 43, 1981.
22. **Tavares, F. C. A.,** Principios do Melhoramento em Função de Parametros Biometricos Aplicados a *Saccharomyces uvarum* Beijerink, Ph.D. thesis, Escola Superior de Agricultura "Luiz de Queiroz", University of São Paulo, Piracicaba, Brazil, 1974.
23. **Allard, R. W.,** *Principles of Plant Breeding*, John Wiley & Sons, New York, 1960.
24. **Boinot, F.,** Le Topinambour dans la fabrication de l'alcool, *Bull. Assoc. Chim.*, 59, 792, 1942.
25. **Guiraud, J. P., Daurelles, J., and Galzy, P.,** Alcohol production from the Jerusalem artichoke using yeasts with inulinase activity, *Biotechnol. Bioeng.*, 23, 1461, 1981.

26. **Margaritis, A. and Bajpal, P.,** Ethanol production from Jerusalem artichoke tubers *(Helianthus tuberosus)* using *Kluyveromyces marxianus* and *Saccharomyces rosei, Biotechnol. Bioeng.*, 24, 941, 1982.
27. **Kosaric, N., Ng, D. S. M., Russell, I., and Stewart, G. G.,** Ethanol production by fermentation: an alternative liquid fuel, *Adv. Appl. Microbiol.*, 26, 147, 1980.
28. **Kierstan, M. P. J.,** Production of fructose syrups from inulin, *Process Biochem.*, 5, 2, 1980.
29. **Echeverrigaray, S. and Tavares, F. C. A.,** Atividade da inulinase em leveduras e fermentações de extratos de *Helianthus tuberosus* L., *Rev. Microbiol. São Paulo*, 16, 127, 1985.
30. **Echeverrigaray, S.,** Estabilidade Genética e Heterose em Híbridos Interespecíficos de Leveduras, M.S. thesis, Escola Superior de Agricultura "Luiz de Queiroz", University of São Paulo, Brazil, 1983.
31. **Stewart, G. G.,** The genetic manipulation of industrial yeast strains, *Can. J. Microbiol.*, 27, 973, 1981.
32. **van Uden, N., Cabeça-Silva, C., Madeira-Lopes, A., and Spencer-Martins, I.,** Selective isolation of derepressed mutants of an α-amylase yeast by the use of 2-de xyglucose, *Biotechnol. Bioeng.*, 22, 651, 1980.
33. **Stewart, G. G., Panchal, C. J., Russell, I., and Sills, A. M.,** Advances in ethanol from sugars and starch: a panoramic paper, in Royal Soc. Canada Int. Symp. on Ethanol from Biomass, Winnepeg, Manitoba, 1982.
34. **Philliskirk, G. and Young, T. W.,** The occurrence of killer character in yeasts of various genera, *Antonie van Leeuwenhoek; J. Microbiol. Serol.*, 41, 147, 1975.
35. **Wickner, R. B.,** Killer of *Saccharomyces cerevisiae:* a double strand ribonucleic acid plasmid, *Bacteriol. Rev.*, 40, 757, 1976.
36. **Gunge, N., Tamaru, A., Ozawa, F., and Sakaguchi, K.,** Isolation and characterization of linear deoxyribonucleic acid plasmids from *Kluyveromuces lactis* and the plasmid-associated killer character, *J. Bacteriol.*, 145, 382, 1981.
37. **Rosini, G.,** The occurrence of killer characters in yeasts, *Can. J. Microbiol.*, 29, 1462, 1983.
38. **Tavares, F. C. A. and Echeverrigaray, S.,** Melhoramento de leveduras floculantes para a produção de etanol, in Anais 6th Simp. Nacional de Fermentação, Sociedade Brasileira de Microbiologia, 1984, 16.
39. **Echeverrigaray, S.,** in preparation.

Chapter 6

PHYSIOLOGY AND GENETICS OF ETHANOL TOLERANCE IN YEAST

Stephen G. Oliver

TABLE OF CONTENTS

I. INTRODUCTION

The use of yeast to ferment sugars to alcohol is probably the oldest of all biotechnological processes and was exploited by the Sumerians and Babylonians to produce beer before 6000 B.C.[1] The role of the organism in leavening bread by the generation of carbon dioxide was developed in Egypt at a much later date, *circa* 4000 B.C., although the ancient Egyptians built on their bread-making expertise in the production of an acidic beer called boza or boozah from a lightly baked "loaf" of germinated grain. The production of more concentrated liquors by distillation was common throughout the known world by the 14th century A.D., having originated in either China or the Middle East.

In spite of its 8000-year history, ethanol produced by biological means has enjoyed only transient popularity as a fuel or chemical feedstock. It reached its peak in Europe immediately prior to and during World War II,[2] but rapidly declined when gasoline became readily available after the war.[3] By the mid 1970s, virtually all industrial ethanol was produced by the catalytic hydration of ethylene with water:[4]

$$C_2H_4 + H_2O \xrightarrow[\substack{300°C \\ 70 \text{ bar pressure} \\ H_3PO_4/SiO_2 \text{ catalyst}}]{} C_2H_5OH$$

This chemical process, unlike the biological one, must be undertaken at high temperature and pressure. Nevertheless, it is still more economically favorable than the fermentation process, provided political or strategic reasons do not dictate the introduction of tax inducements or tariff barriers in favor of the biological route.

Inspection of the cost components of the fermentative production of ethanol shown in Table 1 indicates why it is so difficult to produce ethanol, or many other bulk chemicals, by fermentation on a commercial basis. First, the cost of the carbon substrate represents some two thirds of the cost of the product. Therefore, fluctuations in raw materials prices can render the biological route uncompetitive. While it may be reasonable to predict a continued escalation in oil prices in the long term, the current recession in the world economy has depressed the price of crude oil. Second, although the fermentation process itself occurs at atmospheric pressure and ambient temperatures, there has to be a considerable energy input to distill the ethanol produced from the concentration of 6 to 9% w/v, which is typical of many industrial processes[5] down to 95% w/v. Table 1 demonstrates that the cost of steam for distillation represents about 10% of the costs of the process. This distillation requirement is not only costly in financial terms, but may also threaten the overall energetic efficiency of the process.

The principal biological limitation to the efficient production of ethanol by fermentation is, therefore, that it is produced only in dilute solution.[6] This limitation affects the production of many chemicals by fermentation and the ethanol process may be regarded as a test bed for general solutions to this problem. The reason ethanol is produced at only low concentration is that it is toxic to its producer organism — yeast. In this chapter, I will examine the biochemical and physiological basis of this toxicity, discuss how it is modified by both environmental and genetic factors, and demonstrate how the ethanol tolerance of fermentation may be improved by isolating mutant yeast strains.

Table 1
PROPORTIONAL COSTS OF ETHANOL PRODUCTION FROM MOLASSES

Costs	Percentage
Fixed costs (including labor)	13.6
Materials	5.1
Water	2.7
Electricity	1.4
Steam	9.2
Molasses	68.0

Based on data from Righelato, R. C., *Philos. Trans. R. Soc. Ser. B*, 290, 303, 1980.

II. BIOCHEMICAL AND PHYSIOLOGICAL ASPECTS OF ETHANOL TOLERANCE

A. The Complexity of Ethanol Toxicity and the Definition of "Tolerance"

It has long been recognized that yeast is sensitive to ethanol. For instance, Slator[7] demonstrated that the yield of ethanol became independent of the sugar concentration above a critical value of the latter, and Rahn[8] confirmed that it was not the sugar concentration itself which caused the decrease in the rate of fermentation. A number of methods of defining ethanol tolerance have been advanced over the years. Fermentation rate,[9] glucose consumption,[10] biomass yield,[11] growth rate,[11-13] and cell viability[14] have all been used as indicators of the relative tolerance or sensitivity of various yeast strains to the alcohol. This complexity is not merely apparent, but real. It has been demonstrated that ethanol has different, and separable, effects on the growth rate, fermentation rate, and viability of yeast.[15] When the ethanol tolerance of a particular yeast strain is discussed, therefore, it is essential to specify which physiological parameter has been measured; there can be no universal definition of ethanol tolerance.

While it is generally agreed that the inhibition of yeast growth and fermentation by ethanol is a noncompetitive effect, a variety of kinetic treatments has been used to describe the process. There are three basic types of model in which the degree of inhibition is expressed as either a linear,[12,16,17] hyperbolic,[13,15,18] or exponential[19] function of the ethanol concentration. These three models are expressed in the following equations:

- Linear

$$\mu_i = \mu_{max} - K_i I$$

- Hyperbolic

$$\mu_i = \frac{\mu_{max} \cdot K_i}{K_i - I}$$

- Exponential

$$\mu_i = \mu_{max} \cdot e^{(-K_i I)}$$

where μ_i = specific growth rate of the yeast in the presence of a particular concentration,

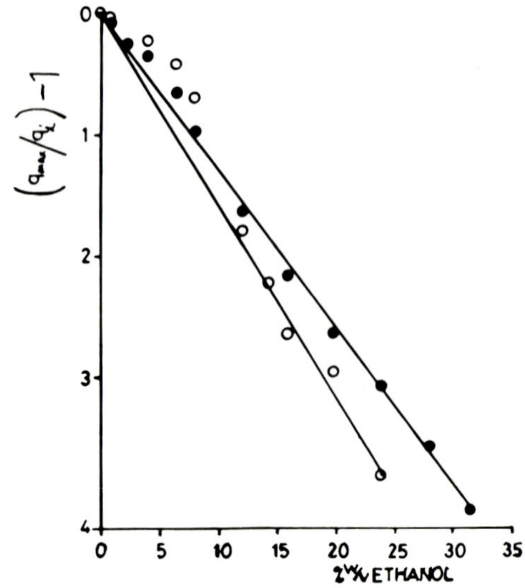

FIGURE 1. Effect of ethanol on fermentation in the two yeast strains: plot for noncompetitive inhibition kinetics. —○—, 5D-cyc; —●—, NCYC479. (Reprinted from Brown, S. W., Oliver, S. G., Harrison, D. E. F., and Righelato, R. C., *Eur. J. Appl. Microbiol.*, 11, 151, 1981. With permission.)

I, of ethanol; μ_{max} = the maximum specific growth rate of the strain in the absence of ethanol; K_i = an inhibition constant, analogous to Km in Michaelis-Menten enzyme kinetics, and represents the value of I at which $\mu_i = 0.5\ \mu_{max}$. All of these equations are suitable for the analysis of ethanol inhibition in the exponential phase of batch culture where substrate is in excess. In order to model substrate-limited continuous culture, more complex equations have been employed which, in essence, incorporate a term for ethanol inhibition into the classic Monod equation.[20-22]

In our own work[15,23] we have followed the lead of Aiba et al.[13] in applying the hyperbolic noncompetitive inhibition model first developed as a general description of product inhibition in microorganisms by Hinshelwood.[24] This treatment is most analogous to the relationship describing noncompetitive inhibition in enzyme kinetics and means that a straight line of slope, K_i, whould be obtained on plotting ($\mu_{max}/\mu_i - 1$) against I. Figure 1 demonstrates that this relationship holds true for the inhibition of fermentation for ethanol. It also shows that a commercial sake yeast is more ethanol tolerant for fermentation than is a laboratory haploid strain, although there is little or no difference between them in terms of the ethanol tolerance of net growth. When growth itself is considered, Figure 2, it is quite obvious that our simple hyperbolic model does not describe the inhibition pattern obtained. The curve for the ethanol inhibition of net growth is, in fact, the product of two inhibitory functions; the inhibition of the true growth rate and the effect of cell death. When the data presented in Figure 2 are corrected for cell viability, the expected straight-line relationship is demonstrated for the alcohol inhibition of the true growth rate (Figure 3). The curve in Figure 3 also reveals that there is a threshold concentration of ethanol (between 1 and 2% w/v) below which there is no observable inhibition of true growth. Holzberg et al.[12] observed a threshold concentration of 2.6 w/v before there was any inhibition of yeast growth on grape juice. These observations emphasize that the simple noncompetitive inhibition model is an imperfect description of the kinetic data.

It was also possible for us to derive an inhibition constant for the effect of ethanol on

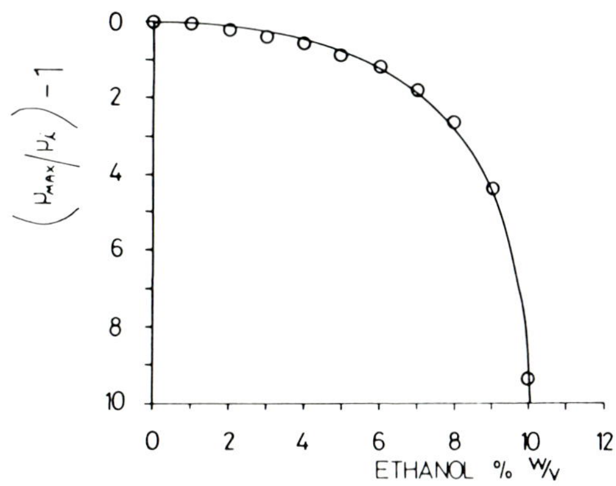

FIGURE 2. Effect of ethanol on growth of 5D-cyc: plot for noncompetitive inhibition kinetics. (Reprinted from Brown, S. W., Oliver, S. G., Harrison, D. E. F., and Righelato, R. C., *Eur. J. Appl. Microbiol.*, 11, 151, 1981. With permission.)

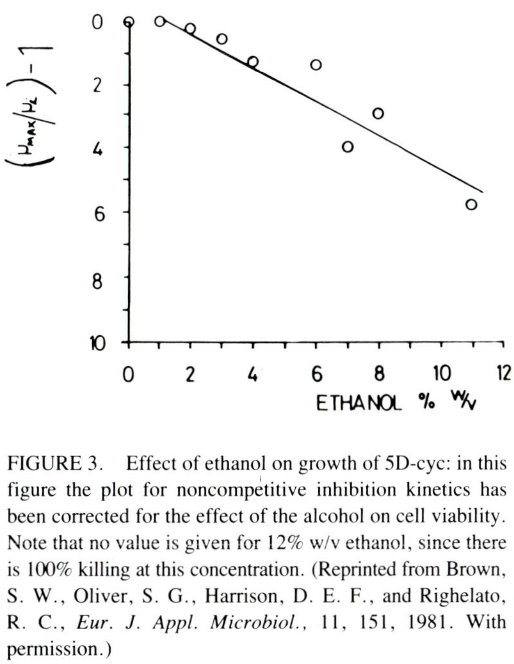

FIGURE 3. Effect of ethanol on growth of 5D-cyc: in this figure the plot for noncompetitive inhibition kinetics has been corrected for the effect of the alcohol on cell viability. Note that no value is given for 12% w/v ethanol, since there is 100% killing at this concentration. (Reprinted from Brown, S. W., Oliver, S. G., Harrison, D. E. F., and Righelato, R. C., *Eur. J. Appl. Microbiol.*, 11, 151, 1981. With permission.)

cell viability. The death rate was determined at each ethanol concentration and it was assumed that the maximum killing effect occurred at an ethanol concentration of 12% w/v, the minimum concentration of the alcohol at which there was no net growth of the culture. This rather pragmatic approach was forced on us by the quantal and irreversible nature of death as a process. Different K_i values were obtained for the inhibition by ethanol of true growth, viability, and fermentation (Table 2). It is noticeable that cell viability is the most sensitive parameter and fermentation the most resistant. This means that yeast cells are still capable of fermentation when dead and, indeed, probably do so in the sequential fed-batch process of the sake fermentation and in ''rapid fermentation'' processes which employ high concentrations of yeast cells.[25]

Table 2
INHIBITION CONSTANTS FOR THE ACTION OF ETHANOL ON *S. UVARUM* 5D-cyc

	K_i (ethanol)	
Physiological parameter	***M***	**(% w/v)**
Growth	0.44	2.01
Viability	0.29	1.32
Fermentation	0.97	4.46

Data from Brown, S. W., Oliver, S. G., Harrison, D. E. F., and Righelato, R. C., *Eur. J. Appl. Microbiol.*, 11, 151, 1981.

Table 3
INHIBITION CONSTANTS FOR THE EFFECT OF ENDOGENOUS AND EXOGENOUS ETHANOL ON CELL GROWTH

Growth temp (°C)	μ_{max} (h^{-1})	K_i (% w/v)	Ref.
	Endogenous Ethanol		
30	0.64	0.52	21
35	0.58	0.5	28
30	0.26	1.37	30
	Exogenous Ethanol		
28	0.31	2.06	26
30	0.43	5.5	27
35	0.64	4.0	20
30	0.31	2.01	15
30	0.41	7.0	30
30	0.34	3.12	30

In our experiments on the kinetics of ethanol inhibition of growth, fermentation, and viability, we employed exogenously added ethanol. This was both an experimental convenience and also permitted us to study ethanol inhibition in the absence of substrate inhibition. The absence of such complicating factors allowed the simple mathematical model described to be applied. However, during the fermentation process itself, ethanol is generated endogenously and so it is important to ascertain whether there is any difference in the inhibitory effect of endogenous and exogenous ethanol.

B. Inhibition by Endogenously Generated or Exogenously Added Alcohol

Hoppe and Hanford[21] have compared the inhibition constants obtained for the effect of ethanol on yeast growth by a number of workers using either exogenously added[20,26,27] or endogenously generated[21,28,29] ethanol. This compilation is given in Table 3, to which I have added some of our own data from experiments with exogenously added ethanol.[15,30] The inhibition constants obtained for the effect of endogenous ethanol on cell growth are uniformly lower than those obtained for exogenous ethanol. While this implies that the endogenously

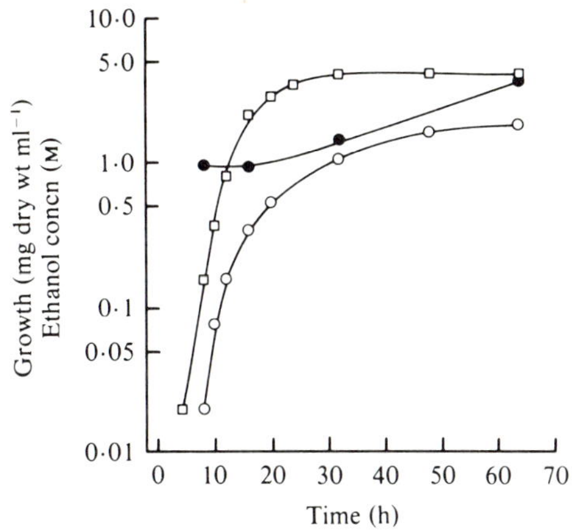

FIGURE 4. Time-course of growth (□) of, and accumulation of intracellular (●) and extracellular (○) ethanol in cultures of *Saccharomyces cerevisiae* NCYC 431. Each value is the mean of three independent determinations. (Reprinted from Beaven, M. J., Charpentier, C., and Rose, A. H., *J. Gen. Microbiol.*, 128, 1447, 1982. With permission.)

generated alcohol is more toxic than that which is exogenously added, these data should be treated with some caution. Not only is there great variation in the yeast strains, culture media, and growth conditions used, but also most workers have neglected to correct their data for the lethal effects of ethanol and thus it is the inhibition of the net growth rate, rather than the true growth rate, which has been measured. What is really needed is some direct measure of the intracellular concentration of ethanol and its relation to the true growth rate of yeast.

Ethanol is usually assumed to cross the yeast cell membrane by simple diffusion,[31] and so the intracellular concentration of ethanol would be expected to reach an equilibrium with that added outside the cell. The same should be true of endogenously generated ethanol and the concentration of expelled ethanol in the culture medium. However, this second equilibrium will depend both on the relative rates of ethanol production and diffusion and whether yeast has some positive means of excreting ethanol against a concentration gradient, as has been suggested by some workers.[32] There are a number of difficulties associated with direct measures of intracellular ethanol content which have not been recognized until recently,[33] namely, that ethanol is rapidly released from yeast cells washed with water or buffer and that the intracellular water content falls during the batch growth cycle. These difficulties were resolved by Beaven and co-workers[33] by an ingenious double-labeling regime which employed tritiated water (which equilibrates with both intra- and extracellular water) and $^{36}Cl^-$ (which is excluded by the yeast cell membrane and thus equilibrates only with interstitial water and the water in the wall). Their results are shown in Figure 4 and demonstrate that ethanol is always excreted from yeast cells down a concentration gradient. It is unfortunate that these workers did not go on to derive a K_i for the effect of intracellular ethanol concentration on the true growth rate.

C. Effects of the Physical and Chemical Environment on Ethanol Tolerance

The study of the kinetics of ethanol inhibition of the growth and metabolism of yeast has revealed it to be a complex phenomenon. This complexity means that the ethanol tolerance of yeast is likely to be subject to considerable modification by both environmental and genetic factors. This section will examine some of the environmental factors.

There has been little systematic study on the effects of medium composition on ethanol tolerance and many workers have employed undefined natural substrates such as grape juice.[12] The impact of sugar concentration on ethanol tolerance has received some attention, however, and it was Gray[34] who first reported that ethanol tolerance decreased with increasing sugar concentration. High substrate concentrations are, of course, essential to obtain a high product concentration, and Nagodawithana et al.[35] observed that cell viability was higher if sugar was added sequentially throughout the fermentation process instead of using a very high initial sugar concentration. This sequential fed-batch regime is exactly the approach which has been used in sake manufacture for centuries[36] and permits the production of a beverage which, at about 15% w/v, has the highest alcohol content of any nondistilled liquor. This exceptional alcohol content cannot be explained solely in terms of the tolerance of production strains of sake yeast, since we found that NCYC 478 (Kyokai No. 7) exhibited greater ethanol tolerance than a laboratory yeast strain, only for fermentation and not for growth or viability[15] (Figure 1). Hayashida and co-workers[37,38] demonstrated that a proteolipid, produced by the *Aspergillus oryzae* which carries out the primary breakdown of starch in the sake process, enhanced the "alcohol durability" of Kyokai No. 7 in anaerobic culture. "Alcohol durability" is a measure of long-term fermentative performance; "alcohol resistability" (or short-term fermentative performance) was unaffected by the presence of the protein-phospholipid complex from the mold. The significance of these results is discussed below in terms of the importance of membrane lipid composition in ethanol tolerance.

The use of high sugar concentrations is also desirable, because high glucose levels repress respiration via the Crabtree effect[39,40] and obviate the need to control rigorously the access of oxygen to the culture. This implies that the Crabtree effect is epistatic to the Pasteur effect, the inhibition of glycolysis by oxygen. Lagunas[41] has argued that the Pasteur effect is irrelevant to yeast metabolism, even at moderate glucose concentrations. Nevertheless, it is certainly possible to demonstrate the induction of respiration by oxygen in continuous culture[42] (Figure 5A), and although Nagodawithana et al.[35] reported that oxygen provided protection against killing by ethanol, it is not clear that this was not a consequence of respiratory activity. The control of the oxygen supply to a yeast fermentation is a critical parameter, since the organism cannot be grown under total anaerobiosis without the addition of a sterol and an unsaturated fatty acid[43,44] (see below).

The effects of temperature on ethanol production and tolerance have been the subject of many studies.[23,35,45-47] Temperature, like ethanol itself, has markedly different effects on net growth and fermentation. In our own studies on *Saccharomyces uvarum* 5D-cyc,[23] we found that the optimum temperature for growth was 35°C, whereas that for fermentation was 40°C. The effect of temperature on the ethanol tolerance of net growth and fermentation also differed. Net growth of *S. uvarum* was most resistant to inhibition by ethanol at temperatures (25 and 30°C) just below the growth optimum (35°C). In contrast, we found that the K_i for ethanol inhibition of fermentation actually increased with increasing temperature.

D. Targets for Ethanol Toxicity

The complexity of the effect of ethanol on the growth and metabolism of yeast suggests that there may be many target sites within the yeast cell for the toxic action of the alcohol. Ethanol is known to be a general denaturant of proteins[48] and, at high concentrations, it reduces the V_{max} of a number of the principal glycolytic enzymes, although some, e.g., phosphofructokinase, are noticeably resistant.[49] That exogenously added ethanol, at least, places yeast cells under stress is witnessed by the fact that ethanol treatment induces them to synthesize the heat shock proteins.[50] A general effect of ethanol treatment is to lower the rate of RNA and protein accumulation in yeast cultures.[82] The inhibition of protein synthesis has been shown to be due to the action of the alcohol at both the initiation and elongation steps of the process[51] ethanol inhibition of translation has been much more extensively studied in mammals using both whole animals and cultured cells (see review[52]).

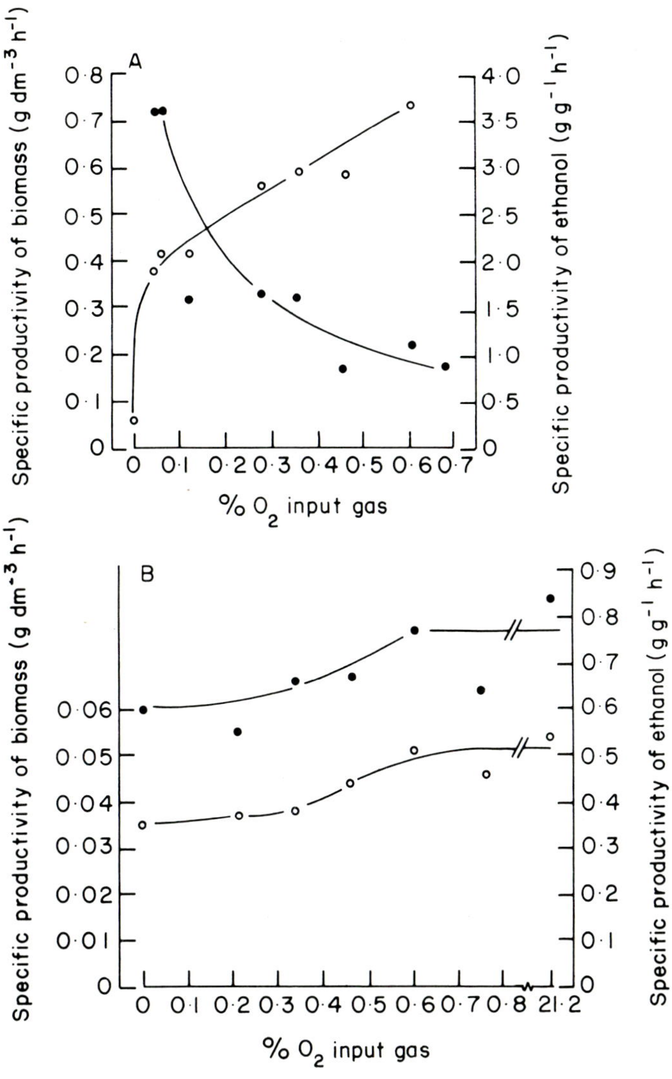

FIGURE 5. Effect of oxygen on growth and ethanol production. (A) 5 D-cyc ρ^+; (B) 5 D-cyc ρ^0. ○—○, specific rate of production of biomass, g dm^{-3}h^{-1}; ●—●, specific rate of production of ethanol, g g^{-1} (dry wt cells) h^{-1}. (Reprinted from Brown, S. W., Sugden, D. A., and Oliver, S. G., *J. Chem. Technol. Biotechnol.*, 34B, 116, 1984. With permission.)

Alcohols are soluble in lipids and the increasing toxicity of higher alkanols for yeast parallels their increasing solubility in lipids.[53] For this reason, much attention has been paid to the plasma membrane as a target site for the toxic action of ethanol.[31] Rose and colleagues have demonstrated that the lipid composition of the yeast cell membrane conditions the susceptibility of cells to the inhibitory effects of ethanol on viability[54] and growth rate.[55] These workers were able to manipulate the lipid composition of yeast membranes by exploiting the organism's requirement for a sterol and an unsaturated fatty acid when growing under conditions of total anaerobiosis.[43,44] They found that membranes enriched in unsaturated fatty acids (e.g., linoleyl residues; C18:2) and sterols with unsaturated side chains (e.g., ergosterol and stigmasterol) increased the resistance of the cells to the killing effects of ethanol as compared with those with membranes enriched in saturated fatty acids (e.g., oleyl residues; C18:1) and sterols with fully saturated side chains (e.g., campesterol and cholesterol). Rose[31] has suggested that the unsaturated residues may increase the fluidity of the yeast plasma membrane and thereby compensate for the decrease in fluidity which occurs on entry of ethanol.[56] These data may also explain the protective effect of the mold proteolipid in sake brewing.[37,38]

A likely effect of any derangement of the yeast plasma membrane by ethanol is that it will interfere with the transport of solutes into the cell. Thomas and Rose[55] noted that the rate of accumulation by yeast of certain amino acids and of glucose was reduced by ethanol and that this effect could be modulated by altering the lipid composition of the membrane. These studies on the effect of ethanol on solute transport have been extended by van Uden and co-workers.[53,57-60] The effect of the alcohol on transport is evidently a nonspecific one, since it inhibits the electroneutral glucose transport system, as well as ammonium transport (an electrogenic uniport) and that of maltose, and by the general amino acid permease (both electrogenic proton symports). The role of ions in solute transport by yeast is discussed by my colleague, A. A. Eddy.[61] Kinetic analyses of the inhibition of these transport systems by ethanol showed that the effect was noncompetitive and followed the exponential pattern described in Equation 3 (above). Inhibition constants calculated for the effect of ethanol on solute transport are comparable to those derived for the effect of exogenously added ethanol on growth rate (see Table 3). For instance, Loureiro-Dias and Peinado[57] gave a K_i of 3% w/v for the inhibition of maltose transport by ethanol.

Leao and van Uden[60] have shown that ethanol may act to inhibit transport by promoting the passive influx of protons into the cell and thus uncoupling electrogenic processes. However, the inhibition of xylose uptake, which enters via the electroneutral glucose transport system, was similar in magnitude to that of maltose or lysine, which enters via proton symports, and so a direct effect of the alcohol on the permease proteins themselves cannot be excluded. Moreover, attention has been focused on the effect of ethanol on the plasma membrane to the exclusion of the mitochondrial and vacuolar membranes. The relative importance of the action of the alcohol at these three membrane sites is not readily assessed by direct biochemical experiments, and it is likely that a genetic approach, using ethanol-sensitive yeast mutants, will be of great utility in identifying the precise biochemical targets of the alcohol toxicity.

III. GENETICAL ASPECTS OF ETHANOL TOLERANCE

Given the pleiotropic nature of the effect of ethanol on yeast, it is most unlikely that any single gene will be solely responsible for the sensitivity or tolerance of the organism to the alcohol. A number of nuclear genes may mutate to confer an ethanol-sensitive phenotype on yeast,[62] and, in our own studies, we have identified the *pep4*-3 mutation[63] as one which confers ethanol sensitivity.[64] The K_i for the ethanol inhibition of the true growth rate is reduced in strain 20B-12 (*pep4*-3) to 3.12% w/v as compared to 7% w/v in its parent X2180-1C *(PEP4)*.[30] The *pep4*-3 mutation is a nuclear point mutation which reduces the activity of a broad spectrum of vacuolar hydrolases in yeast, including protease A, protease B, carboxypeptidase,[63] alkaline phosphatase,[65] and ribonuclease.[66] We found that the ethanol-sensitive phenotype conferred by this mutation was temperature conditional, in that it was only observed at above 25°C,[64] and suggested that this might reflect the temperature shift involved in membrane phase transitions.[67] If this is so, then it must be the vacuolar membrane which is involved in ethanol sensitivity in this case, since *pep4*-3 has no effect on the transport of enzymes through the plasma membrane into the periplasm.[66]

A. The Cytoplasmic Petite Mutation and Ethanol Tolerance

It is not just nuclear mutations that may affect the ethanol tolerance of yeast, but cytoplasmic ones as well. Bacila and co-workers[68,69] have claimed that cytoplasmic petite mutants are more efficient than their grande (rho+) parents in the production of ethanol by fermentation. This is an attractive idea, since petite mutants are incapable of respiratory metabolism and so are not subject to the Pasteur effect (O_2 suppression of fermentation).

The relative performance of grande and petite cells in ethanol production has been evaluated in three laboratories.[42,70,71] Alexander and Detroy[71] showed that the ethanol yield was similar from cultures of either grande or petite cells. However, the supply of oxygen to their cultures was not carefully regulated. In both our own study[42] and that of Esser and co-workers,[70] it was found that neither the yield nor productivity of ethanol by petite cultures exceeded that of a grande strain when the latter was grown under conditions in which the supply of oxygen was rigorously controlled. We found that the rate of production of both cells and ethanol increased with increasing oxygen, in cultures of a petite mutant, up to an oxygen level of 0.6% of the gas feed[42] (Figure 5B).

Between 0.6 and 21.2% oxygen in the input gas flow, there was no further increase in the specific rate of production of either biomass or ethanol. It was demonstrated that the culture was limited by neither oxygen nor glucose at these oxygen levels, and we reasoned that growth and fermentation were limited by product inhibition. On examining the effect of ethanol on the growth rate of grande and petite cells, we found that the latter had a reduced ethanol tolerance. This may have been the result of some change in the yeast cell surface induced by the cytoplasmic petite mutation, since Wilkie and co-workers[72-74] have found that a number of surface phenomena, such as flocculation and solute transport, are altered in cytoplasmic, but not nuclear, petite yeast mutants. Nuclear petites have a more specific lesion in their respiratory chain than the cytoplasmic mutants and so may be of greater practical value in circumventing the Pasteur effect and the need to control carefully the supply of oxygen to the fermentation.

B. The Isolation of Ethanol-Tolerant Yeast Mutants by Feedback Continuous Selection

The identification of nuclear and cytoplasmic mutations which confer an ethanol-sensitive phenotype in yeast may make a valuable contribution to our understanding of the biochemical targets for ethanol toxicity, but it makes no direct contribution to increasing the concentration of ethanol which may be obtained by fermentation. To achieve this, ethanol-tolerant mutants are required, and attempts to isolate them by conventional agar plate screening procedures have failed.[75,76] This failure is not surprising in view of the pleiotropic nature of ethanol toxicity in yeast. It is likely that the mutation of a number of genes will be required to improve the ethanol tolerance of yeast and, moreover, such improvements are likely to be small, since *Saccharomyces cerevisiae* is already a highly tolerant organism.

In this situation, when only small quantitative increases are likely to be obtained as a result of multiple (rather than single) mutations, the use of continuous selection is indicated.[77,78] However, the very complexity of the inhibitory effects of ethanol on yeast makes it difficult to design a suitable selection regime. For this reason we adopted a system in which the intensity of selection was determined by the culture itself via a feedback control circuit.[79]

The feedback selection system employed is diagrammed in Figure 6. An IR analyzer was used to monitor continuously the fermentation activity of the culture by determining the concentration of carbon dioxide in the exit gas. When this concentration equaled or exceeded the value set on the potentiometric controller, a relay closed. This switched on a peristaltic pump and introduced a concentrated ethanol solution into the culture vessel. When addition of the alcohol had reduced the carbon dioxide concentration of the exit gas to below the value set by the controller, the relay opened and the ethanol pump was switched off. The ethanol was then gradually diluted out of the fermenter vessel and the rate of carbon dioxide production increased. When the carbon dioxide concentration in the exit gas exceeded the

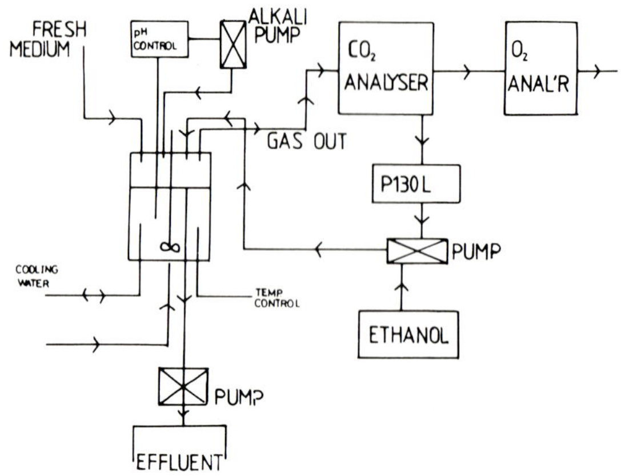

FIGURE 6. Continuous culture apparatus for feedback selection. (Reprinted from Brown, S. W. and Oliver, S. G., *Eur. J. Appl. Microbiol. Biotechnol.*, 16, 119, 1982. With permission.)

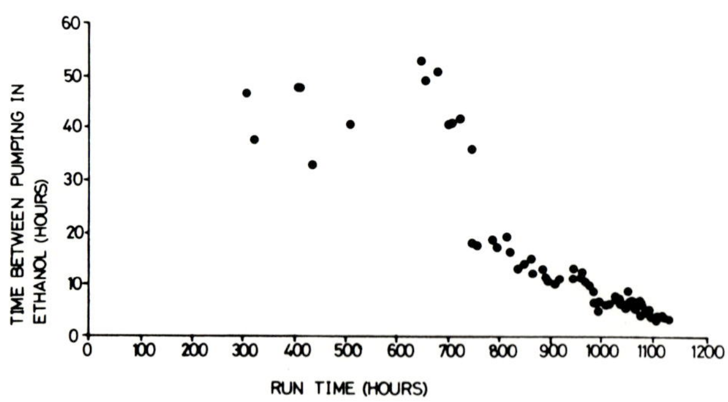

FIGURE 7. Improvement in ethanol tolerance of culture. The graph describes the frequency of switching of the ethanol pump in response to an increase in CO_2 concentration of the exit gas. Operation of the pump was detected by the small, downward "blip" on the trace from the CO_2 analyzer which was produced each time ethanol first reached the culture. Each point represents a 7-hr moving average of the interval between operations of this pump. (Reprinted from Brown, S. W. and Oliver, S. G., *Eur. J. Appl. Microbiol. Biotechnol.*, 16, 119, 1982. With permission.)

value set by the controller once more, the pump was switched on again. Therefore, the system was functionally analogous to a turbidostat. However, it was the rate of carbon dioxide production for the culture, rather than its turbidity, which was held constant, and the supply of inhibitor, rather than nutrients, which was regulatory.

The system indicated an improvement in the ethanol tolerance of the culture by an increase in the frequency with which the ethanol pump was switched on by the control system. The pump did not switch on at all for the first 12 days of the experiment. After that, it switched on occasionally until, 27 days (650 hr) into the experiment, there was a dramatic increase in the frequency with which the pump switched on (Figure 7). This increase in switching coincided with a number of changes in the physiological state of the culture (Figure 8). The specific rate of carbon dioxide production and the concentration of ethanol in the growth medium increased, as did the viability of the culture. These changes were accompanied by

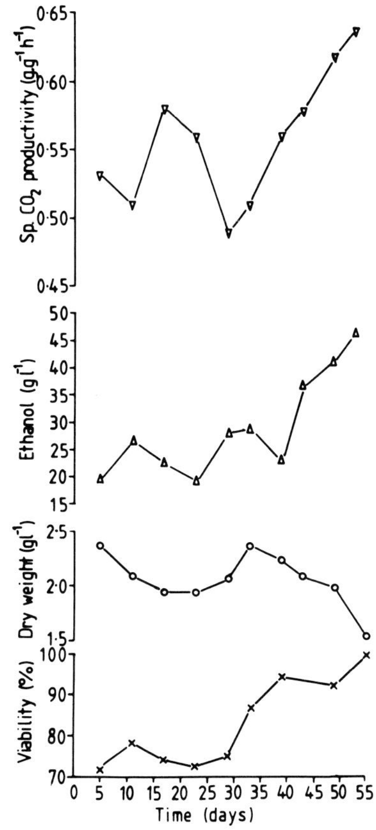

FIGURE 8. Changes in the performance of the culture during the selection experiment. x—x, cell viability; o—o, dry weight; △—△, ethanol concentration; ▽—▽, specific rate of production of CO_2. These parameters were measured every 2 days; the graph shows the arithmetic mean values for successive 6-day periods. (Reprinted from Brown, S. W. and Oliver, S. G., *Eur. J. Appl. Microbiol. Biotechnol.*, 16, 119, 1982. With permission.)

a fall in the concentration of biomass in the culture vessel, as predicted by analogy with a turbidostat. The abruptness of these changes might be interpreted as being the result of some particular mutational event. This seems unlikely; mutants with a range of phenotypes were obtained and genetic analysis showed that these carried mutations in more than one gene.

In the first instance, mutants were isolated by selecting cells which formed colonies on 12% (w/v) ethanol plates. This concentration of alcohol was lethal to the parent strain used in the experiment. The fermentation performance of these mutants was subsequently investigated and the best were found to ferment at twice the rate of the parent strain in the presence of 10% (w/v) ethanol (Table 4). The mutants fell into two classes: some, like SB159 and SB160[80] (Table 4), were selected on the basis of an enhanced overall fermentation rate, the K_i of ethanol inhibition of fermentation being little changed from that in the wild type. Others, such as SB154 and SB155, had an overall fermentation rate no better than that of the wild-type strain, but the K_i of ethanol for fermentation inhibition was significantly increased. Mutants of the first class may find immediate application in the commercial production of ethanol. Indeed, SB160 has been shown to reduce significantly attentuation

Table 4
COMPARISON OF THE FERMENTATION PERFORMANCE OF MUTANTS OF *SACCHAROMYCES UVARUM* 5D-cyc ISOLATED BY FEEDBACK CONTINUOUS SELECTION[80]

Strain	Fermentation rate (QCO_2; $gg^{-1}h^{-1}$)	K_i (fermentation) (% w/v ethanol)
5D-cyc (wild-type)	24.69	4.37
SB154	20.40	8.22
SB155	19.11	5.13
SB159	30.51	4.58
SB160	43.79	4.56

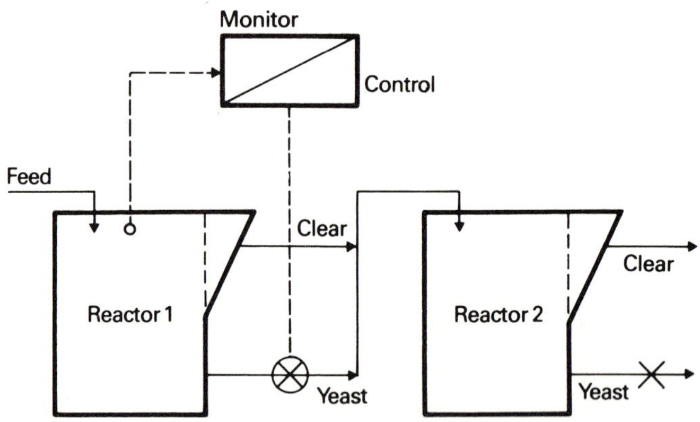

FIGURE 9. A schematic representation of two reactors, each with separate effluents for biomass suspension and biomass-free streams, with control of the biomass effluent from the first reactor in response to monitoring of the first reactor kinetics. (Reprinted from Bu'lock, J. D., Comberbach, M. C., Ghommidh, C., and Williams, P. D., *Chem. Ind.*, 12, 432, 1984. With permission.)

times in batch fermentation on a pilot-plant scale. However, it is likely that mutants of the second class will be more valuable in the long run if they are used as starting material for further strain development.

It should be noted that although the mutants isolated could survive and ferment at enhanced rates in the presence of high concentrations of alcohol, the selection regime never exposed the culture to more than 5% (w/v) ethanol. This illustrates the advantage of allowing the culture to determine the intensity of selection via the feedback system. If we had imposed the selection, we would certainly have used ethanol concentrations higher than 5% (w/v).

IV. APPLICATIONS TO INDUSTRIAL-SCALE PROCESSES

A number of the findings discussed have relevance for industrial-scale production of ethanol by yeast fermentations. The most striking feature of the physiological data is the different ways in which fermentation and growth react to ethanol toxicity.[15] These differences have been exploited by Bu'Lock[81] at the University of Manchester in a two-stage continuous process. In this system (Figure 9), two continuous-flow reactors are linked in series. In the first reactor, yeasts grow and produce ethanol to a moderate level; they are then transferred to a second reactor where they continue to produce ethanol as they die. The data on temperature effects on ethanol tolerance (see above) indicate that the system could be further improved by running the first reactor at temperatures (e.g., 25°C) below optimum in order to enhance the ethanol tolerance for growth, whereas the second reactor should be run at a higher temperature (e.g., 40°C) which is optimal for fermentation. The use of yeasts which are nuclear petites and have been selected for ethanol tolerance by feedback continuous selection should both simplify the control systems and improve the final concentration of ethanol obtained.

ACKNOWLEDGMENTS

In writing this chapter, I have drawn heavily on the thesis research of my two former postgraduate students, Stephen W. Brown and David A. Sugden, with whom I have enjoyed many stimulating discussions. S. W. Brown was supported by a CASE award from the Science and Engineering Research Council and Tate and Lyle plc. D. A. Sugden was supported by a similar award from Bass Trade Development Ltd. I have also benefited from the advice and encouragement of David Harrison and Renton Righelato.

REFERENCES

1. **Corran, H. S.,** *A History of Brewing,* David and Charles, London, 1975.
2. **Kosaric, N., Ng, D. C. M., Russell, I., and Stewart, G. G.,** Ethanol production and fermentation: an alternative liquid fuel, *Adv. Appl. Microbiol.,* 26, 147, 1980.
3. **Freeman, J. H., Appeldoorn, J. K., Berger, J. E., Koehl, W. S., Lindquist, R. H., Taliaferro, H. R., and Gray, D. S.,** American Petroleum Institute, Publ. No. 4261, July 1976.
4. **Miller, D. L.,** Ethanol fermentation and potential, *Biotechnol. Bioeng. Symp.,* No. 5, 345, 1975.
5. **Righelato, R. C.,** Anaerobic fermentation: alcohol production, *Philos. Trans. R. Soc. Ser. B,* 290, 303, 1980.
6. **Oliver, S. G.,** Biological limits to ethanol production, *Chem. Ind.,* 12, 425, 1984.
7. **Slator, A.,** Studies in fermentation. I. The chemical dynamics of alcoholic fermentation by yeast, *J. Chem. Soc. (London),* 89, 128, 1906.
8. **Rahn, O.,** The decreasing rate of fermentation, *J. Bacteriol.,* 18, 207, 1929.
9. **Rubner, M.,** The physiology and nutrition of yeast during alcoholic fermentation, *Arch. Physiol. (Suppl.),* 1, 1912.
10. **Gray, W. D.,** Studies on the alcohol tolerance of yeasts, *J. Bacteriol.,* 42, 561, 1941.
11. **Troyer, J. R.,** A relation between cell multiplication and alcohol tolerance in yeasts, *Mycologia,* 45, 20, 1953.
12. **Holzberg, I., Finn, R. K., and Steinkraus, K. H.,** A kinetic study of the alcoholic fermentation of grape juice, *Biotechnol. Bioeng.,* 9, 413, 1967.
13. **Aiba, S., Shoda, M., and Nagatani, M.,** Kinetics of product inhibition in alcohol fermentation, *Biotechnol. Bioeng.,* 10, 845, 1968.
14. **Thomas, D. S., Hossack, J. A., and Rose, A. H.,** Plasma-membrane lipid composition and ethanol tolerance in *Saccharomyces cerevisiae, Arch. Microbiol.,* 117, 239, 1978.
15. **Brown, S. W., Oliver, S. G., Harrison, D. E. F., and Righelato, R. C.,** *Eur. J. Appl. Microbiol.,* 11, 151, 1981.
16. **Lee, J. H., Williamson, D., and Rogers, P. L.,** The effect of temperature on the kinetics of ethanol production by *Saccharomyces uvarum, Biotechnol. Lett.,* 2, 83, 1980.
17. **Ghose, T. K. and Tyagi, R. D.,** Rapid ethanol fermentation of cellulose hydrosylate. II. Product and substrate inhibition and optimisation of fermenter design, *Biotechnol. Bioeng.,* 21, 1401, 1979.
18. **Novak, M., Strehaiano, P., Moreno, M., and Goma, G.,** Alcoholic fermentation: on the inhibitory effect of ethanol, *Biotechnol. Bioeng.,* 23, 20, 1981.
19. **Moulin, G., Boze, H., and Galzy, P.,** Inhibition of alcoholic fermentation by substrate and ethanol, *Biotechnol. Bioeng.,* 22, 2375, 1980.

20. **Bazua, C. D. and Wilke, C. R.,** Ethanol effects on the kinetics of a continuous fermentation with *Saccharomyces cerevisiae, Biotechnol. Bioeng. Symp.*, 7, 105, 1977.
21. **Hoppe, G. K. and Hanford, G. S.,** Ethanol inhibition of anaerobic yeast growth, *Biotechnol. Lett.*, 4, 39, 1982.
22. **Righelato, R. C., Rose, D., and Westwood, A. W.,** Kinetics of ethanol production by yeast in continuous culture, *Biotechnol. Lett.*, 3, 3, 1981.
23. **Brown, S. W. and Oliver, S. G.,** The effect of temperature on the ethanol tolerance of the yeast, *Saccharomyces uvarum, Biotechnol. Lett.*, 4, 269, 1982.
24. **Hinshelwood, C. N.,** *The Chemical Kinetics of the Bacterial Cell,* University Press, Oxford, 1946.
25. **Nagodawithana, T. W. and Steinkraus, K. H.,** Influence of the rate of ethanol production and accumulation of the viability of *Saccharomyces cerevisiae* in 'rapid fermentation', *Appl. Environ. Microbiol.*, 31, 158, 1976.
26. **Egamberdiev, N. B. and Ierulasimskii, N. D.,** Effect of ethanol concentration on the growth rate of *Saccharomyces vini, Mikrobiologiya,* 37, 689, 1968.
27. **Aiba, S. and Shoda, M.,** Reassessment of product inhibition in alcohol fermentation, *J. Fermentation Technol.*, 47, 790, 1969.
28. **Cysewski, G. R.,** Fermentation Kinetics and Process Economics for the Production of Ethanol, Ph.D. thesis, University of California at Berkeley, 1976.
29. **Pironti, F. F.,** Kinetics of Alcohol Fermentation, Ph.D. thesis, Cornell University, Ithaca, N.Y., 1971.
30. **Sugden, D. A.,** Molecular and Physiological Studies on Yeast and Fungal Protease Genes and Their Mutants, Ph.D. thesis, University of Manchester, England, 1985.
31. **Rose, A. H. and Beaven, M. J.,** End-product tolerance and ethanol, in *Trends in the Biology of Fermentation for Fuels and Chemicals,* Hollaender, A., Rabson, R., San Pietro, A., Valentine, R., and Wolfe, R., Eds., Plenum Press, New York, 1981, 513.
32. **Goma, G., Moreno, M., and Strehaiano, P.,** Mechanism of inhibition during alcohol fermentation in strict anaerobiosis, in *Proc. Symp. on Bioconversion and Bioengineering,* Ghose, T. K., Ed., Indian Institute of Technology, New Delhi, 1981, 97.
33. **Beaven, M. J., Charpentier, C., and Rose, A. H.,** Production and tolerance of ethanol in relation to phospholipid fatty-acyl composition in *Saccharomyces cerevisiae* NCYC431, *J. Gen. Microbiol.*, 128, 1447, 1982.
34. **Gray, W. D.,** The sugar tolerance of four strains of distillers yeast, *J. Bacteriol.*, 49, 445, 1945.
35. **Nagodawithana, T. W., Castellano, C., and Steinkraus, K. H.,** Effect of dissolved oxygen, temperature, initial cell count and sugar concentration on the viability of *Saccharomyces cerevisiae* in rapid fermentation, *Appl. Microbiol.*, 28, 383, 1974.
36. **Kodama, K.,** *The Yeasts,* Vol. 3, Rose, A. H., Harrison, J. S., Eds., Academic Press, London, 1970, 225.
37. **Hayashida, S., Der Fong, D., and Hongo, M.,** Mechanism of formation of high concentration alcohol in sake brewing. IX. Function of the high concentration alcohol-producing factor, *Agric. Biol. Chem.*, 38, 2001, 1974.
38. **Hayashida, S., Der Fong, D., and Hongo, M.,** Mechanism of formation of high concentration alcohol in sake brewing. X. Physiological properties of yeast cells grown in the proteolipid-supplemented medium, *Agric. Biol. Chem.*, 39, 1025, 1975.
39. **Crabtree, H. G.,** Observations on the carbohydrate metabolism of tumours, *Biochem. J.*, 23, 536, 1928.
40. **Fiechter, A., Fuhrmann, G. F., and Kapelli, O.,** Regulation of glucose metabolism in growing yeast cells, *Adv. Microb. Physiol.*, 22, 123, 1982.
41. **Lagunas, R.,** Energetic irrelevance of aerobiosis for *S. cerevisiae* growing on sugars, *Mol. Cell. Biochem.*, 27, 39, 1979.
42. **Brown, S. W., Sugden, D. A., and Oliver, S. G.,** Ethanol production and tolerance in grande and petite yeast, *J. Chem. Technol. Biotechnol.*, 34B, 116, 1984.
43. **Andreasen, A. A. and Stier, T. J. B.,** Anaerobic nutrition of *Saccharomyces cerevisiae.* I. Ergosterol requirement for growth in defined medium, *J. Cell Comp. Physiol.*, 41, 23, 1953.
44. **Andreasen, A. A. and Stier, T. J. B.,** Anaerobic nutrition of *Saccharomyces cerevisiae.* II. Unsaturated fatty acid requirement for growth in defined medium, *J. Cell Comp. Physiol.*, 43, 271, 1954.
45. **Merrit, N. R.,** Influence of temperature on some properties of yeast, *J. Inst. Brewing,* 72, 374, 1966.
46. **Krouwel, P. G. and Braber, L.,** Ethanol production by yeast at supraoptimal temperatures, *Biotechnol. Lett.*, 1, 403, 1979.
47. **Lee, K. J., Williamson, D., and Rogers, P. L.,** *Biotechnol. Lett.*, 2, 83, 1980.
48. **Cohn, E. J., Gurd, F. R. N., Sungenor, D. M., Barnes, B. A., Brown, R. K., Derouaux, G., Gillespie, J. M., Kohnt, R. W., Lever, W. F., Liu, C. H., Mittelman, D., Moutin, R. F., Schmid, K., and Uroma, E.,** A system for the separation of the components of human blood: quantitative procedures for the separation of the protein components of human plasma, *J. Am. Chem. Soc.*, 72, 465, 1950.

49. **Nagodawithana, T. W., Whitt, J. T., and Cutaia, A. J.,** Study of the feedback effect of ethanol on selected enzymes of the glycolytic pathway, *J. Am. Soc. Brewing Chem.*, 35, 179, 1977.
50. **Plesset, J., Palm, C., and McLaughlin, C. S.,** Induction of heat shock proteins and thermotolerance by ethanol in *Saccharomyces cerevisiae*, *Biochem. Biophys. Res. Commun.*, 108, 1340, 1982.
51. **Swedes, J., Plesset, J., and McLaughlin, C. S.,** Effects of ethanol on protein synthesis in *Saccharomyces cerevisiae*, submitted for publication.
52. **Noble, E. P. and Tewari, S.,** Protein and ribonucleic acid metabolism in brains of mice following chronic alcohol consumption, *Ann. N.Y. Acad. Sci.*, 215, 333, 1973.
53. **Leao, C. and van Uden, N.,** Effects of ethanol and other alkanols on the glucose transport system of *Saccharomyces cerevisiae*, *Biotechnol. Bioeng.*, 24, 2601, 1982.
54. **Hossack, J. A. and Rose, A. H.,** Fragility of plasma membranes in *Saccharomyces cerevisiae* enriched in different sterols, *J. Bacteriol.*, 127, 67, 1976.
55. **Thomas, D. S. and Rose, A. H.,** Inhibitory effect of ethanol on growth and solute accumulation by *Saccharomyces cerevisiae* as affected by plasma-membrane lipid composition, *Arch. Microbiol.*, 122, 49, 1979.
56. **Ingram, L. O.,** Adaptation of membrane lipids to alcohols, *J. Bacteriol.*, 125, 670, 1976.
57. **Loureiro-Dias, M. C. and Peinado, J. M.,** Effect of ethanol and other alkanols on the maltose transport system of *Saccharomyces cerevisiae*, *Biotechnol. Lett.*, 4, 721, 1982.
58. **Leao, C. and van Uden, N.,** Effects of ethanol and other alkanols on the ammonium transport system of *Saccharomyces cerevisiae*, *Biotechnol. Bioeng.*, 25, 2085, 1983.
59. **Leao, C. and van Uden, N.,** Effects of ethanol and other alkanols on the general amino acid permease of *Saccharomyces cerevisiae*, *Biotechnol. Bioeng.*, 26, 403, 1984.
60. **Leao, C. and van Uden, N.,** Effects of ethanol and other alkanols on passive proton influx in the yeast *Saccharomyces cerevisiae*, *Biochim. Biophys. Acta*, 774, 43, 1984.
61. **Eddy, A. A.,** Mechanisms of solute transport in selected eukaryotic microorganisms, *Adv. Microb. Physiol.*, 23, 1, 1982.
62. **Aguilera, A., del Castillo, L., and Benitez, T.,** Alcohol and sucrose tolerant wine yeast strains, *Cien. Biol. (Portugal)*, 7, 89, 1982.
63. **Jones, E. W.,** Proteinase mutants of *Saccharomyces cerevisiae*, *Genetics*, 85, 23, 1977.
64. **Sugden, D. A. and Oliver, S. G.,** Reduced ethanol tolerance: one of the pleiotropic effects of the *pep 4*.3 mutation in *Saccharomyces cerevisiae*, *Biotechnol. Lett.*, 5, 419, 1983.
65. **Kaneko, Y., Toh-E, A., and Oshima, Y.,** Identification of the genetic locus for the structural gene and a new regulatory gene for the synthesis of repressible alkaline phosphatase in *Saccharomyces cerevisiae*, *Mol. Cell Biol.*, 2, 127, 1982.
66. **Jones, E. W., Zubenko, G. S., Parker, R. R., Hemmings, B. A., and Hasilik, A.,** Pleiotropic mutations in *Saccharomyces cerevisiae* which cause deficiencies for proteases and other vacuolar enzymes, in *Alfred Benzon Symp. 16: Molecular Genetics of Yeast*, Von Wettstein, D., Stenderup, A., Kielland-Brandt, M., and Friis, J., Eds., Munksgaard, Copenhagen, 1981, 182.
67. **Melchior, D. L.,** Lipid phase transitions and the regulation of membrane fluidity in prokaryotes, *Curr. Top. Membranes Transport*, 17, 263, 1982.
68. **Bacila, M. and Horii, J.,** Improving the efficiency of alcohol production with respiration-deficient yeasts, *Trends Biochem. Sci.*, 4, 59, 1979.
69. **Bacila, M., Xavier, A. W., and Horii, M.,** Induction of respiration-deficient mutants of *Saccharomyces* and evaluation of their efficiency for ethanol production, in *Biochemistry and Genetics of Yeasts*, Bacila, M., Horecker, B. L., and Stoppani, A. O. M., Eds., Academic Press, New York, 1978, 577.
70. **Esser, K., Schmidt, U., and Stahl, U.,** Ethanol and biomass production of wild strains and respiratory deficient mutants of *Saccharomyces cerevisiae* under anaerobic and aerobic conditions, *Eur. J. Appl. Microbiol. Biotechnol.*, 16, 161, 1982.
71. **Alexander, N. J. and Detroy, R. W.,** Alcohol production by mitochondrial mutants and wild type *Saccharomyces cerevisiae*, *Biotechnol. Lett.*, 5, 165, 1983.
72. **Evans, I. H., Diala, E. S., Earl, A., and Wilkie, D.,** Mitochondrial control of cell surface characteristics in *Saccharomyces cerevisiae*, *Biochim. Biophys. Acta*, 602, 201, 1980.
73. **Evans, I. H. and Wilkie, D.,** Mitochondrial role in the induction of sugar utilisation in *Saccharomyces cerevisiae*, *Genet. Res.*, 32, 171, 1978.
74. **Linstead, D., Evans, I. H., and Wilkie, D.,** Mitochondrial genetic factors in the cellular response to chlorimipramine in *Saccharomyces cerevisiae*, in *Biogenesis of Mitochondria*, Kroon, A. M. and Saccone, C., Eds., Academic Press, New York, 1974, 179.
75. **Ismail, A. A. and Ali, A. M. M.,** Selection of high ethanol-yielding *Saccharomyces*. I. Ethanol tolerance and the effect of training in *Saccharomyces cerevisiae* Hansen, *Folia Microbiol.*, 16, 346, 1971.
76. **Ismail, A. A. and Ali, A. M. M.,** Selection of high ethanol-yielding *Saccharomyces*. II. Genetics of ethanol tolerance, *Folia Microbiol.*, 16, 350, 1971.

77. **Harder, W., Kuenen, J. G., and Matin, A. A.,** Microbial selection in continuous culture, *J. Appl. Bacteriol.*, 43, 1, 1977.
78. **Dykhuizen, D. E. and Hartl, D. L.,** Selection in chemostats, *Microbiol. Rev.*, 47, 150, 1983.
79. **Brown, S. W. and Oliver, S. G.,** Isolation of ethanol-tolerant mutants of yeast by continuous selection, *Eur. J. Appl. Microbiol. Biotechnol.*, 16, 119, 1982.
80. **Brown, S. W.,** Ethanol Tolerance in the Yeast *Saccharomyces*, Ph.D. thesis, University of Kent, Canterbury, 1983.
81. **Bu'lock, J. D., Comberbach, M. C., Ghommidh, C., and Williams, P. D.,** A computer-controlled fermentation operating system for high-level alcohol production, *Chem. Ind.*, 12, 432, 1984.
82. **Stephens, J. and Oliver, S. G.,** unpublished observations.

Chapter 7

YEASTS IN BIOTECHNOLOGY

Seymour Fogel and Juliet W. Welch

TABLE OF CONTENTS

I. INTRODUCTION

Contrasted with other principal microorganism groups — algae, bacteria, protozoa, and fungi — yeasts are represented by comparatively few genera and species. Thus, about 350 to 500 species fall into 40 genera, compared to tens of thousands of species in the algae, bacteria, and protozoa. Nonphotosynthetic and lacking a pliable, flexible cell wall, yeasts are readily delimited from the algae and protozoa and their larger average unicellular size readily distinguishes them from most bacilli, cocci, and spirilla.

Primarily, it is the single cell habit that distinguishes yeast from other fungi. This single cell habit confers several unique advantages compared to a mycelial growth habit. Thus, weight for weight, yeasts display a much higher metabolic rate — probably a reflection of the greater surface area/volume ratio. Enzymologically powerful and varied, yeasts are also widely adapted to a range of diverse ecological niches similar to those exploited by bacteria.

Rapid and efficient in the conversion of sugar to CO_2 and alcohol, the yeasts of practical value are long known from empirical arts such as distilling, baking, brewing, and wine making. A model brewery is recorded at Thebes from 2000 B.C. or the 11th Dynasty. The salient industrial applications of yeasts are baking, human and animal feed, and ethanol production for potable, power, and synthetic uses. The importance of the organism is easily assessed from the fact that 20 years ago, the annual world-wide yeast production amounted to over a million tons. Moreover, an appreciable fraction of the world fish and rice eating population has begun to shift to bread, chicken, and meat. Thus, the annual yeast production has doubled or tripled and can be expected to increase.

II. CONTRIBUTIONS OF YEAST TO MODERN BIOTECHNOLOGY

Clearly, to summarize and encapsulate the substantive content of this milestone conference* is probably beyond the ken and scope of any individual scientist. Our knowledge of yeast has proliferated and expanded at a rate and in directions so varied that a barely adequate statement concerning this single organism would require at least three volumes — each some 600 to 700 pages in length. The Renaissance yeast biologist's day has passed. Nonetheless, "it can be concluded that the competitive application of yeasts in modern biotechnology brings into play a wide range of scientific disciplines and relies on highly developed and quantitatively precise techniques for research and process control."

In this sketchy overview, we would like to consider some preliminary studies in several areas. Probably, these foreshadow fertile arenas that are as yet largely unexploited. First, a few remarks on the oldest approach, i.e., strain improvement by breeding and selection. See excellent summaries and reviews by Spencer,[1] and also Stewart,[2] Johnson and Oberman,[3] and von Wettstein et al.[4] Conventional cell hybridization remains a powerful and useful tool in the development of industrial yeast strains. For example, in the study of Anderson and Martin,[5] among a large number of hybrids only 2% were more flocculent than the parental brewing strains and 13% were less flocculent. In a separate study by Reader,[6] strains with accelerated attenuation rates or higher mass yields were obtained. However, mating segregants are often obtained from industrial strains only with considerable effort. Polyploidy, aneuploidy, structural rearrangements, and low frequency sporulation or high spore inviability collectively comprise imposing obstacles to any breeding program.

In consequence, the approach provided by protoplast fusion attracted the attention of serious industrial yeast geneticists. Moreover, the possibility of achieving gene transfers and hybrids between sexually isolated organisms representing different genera and species seemed

* The Biochemistry and Molecular Biology of Industrial Yeasts, Brook Lodge, Kalamazoo, Mich., October 25 to 30, 1985.

especially promising. The production of interspecific and intergeneric hybrids by protoplast fusion has been reported by several different authors. However, the fusion product was both difficult to attain and often quite unstable. Studies reported by Spencer et al.[7] are especially illustrative. These researchers were able to generate relatively stable hybrids in high frequency between several yeast species, i.e., between *Candida pseudotropicalis, Schizosaccharomyces pombe, Saccharomyces rosei, Hansenula capsulata* as one parent, and a petite mutant of *S. diastaticus* as the second parent. Stability was attributed to the uniparental mitochondrial contribution. The hybrid sporulated abundantly, and though the spores resembled the *S. diastaticus* parent, the asci of the hybrid were morphologically distinctive. Composition of the maintenance medium altered the vegetative stability of the hybrid. Thus, cultures maintaining glycerol as the carbon source lost their ability to sporulate after a few transfers, and the cell morphology more nearly resembled the *Hansenula* parent.

Typically, fusions involve protoplasts prepared with a bacterial, fungal, or snail enzyme preparation. Before digestion, cells are treated in a Tris-HCl buffer containing EDTA, mercaptoethanol, and 0.6 to 0.8 *M* KCl or sorbitol as an osmotic stabilizer. Fusions are affected in PEG 6000 containing calcium ions and then plated as an agar overlay on osmotically balanced YEPD agar. Recent technical developments are summarized by Ferenczy[8] and an electrophoretic fusion method has been reported by Weber and others.[9]

As with conventional cell mating, protoplast fusions conjoin entire genomes. In turn, this generates a complex problem of selecting out the desired product by subsequent meiotic or parasexual cycles.

Other approaches, such as vectored or nonvectored DNA transformation or interrupted matings involving the karyogamy-defective *kar1* mutant, first reported by Conde and Fink,[10] offer distinctive possibilities with limited genetic transfers. Thus, only single genes or a few genes can be conveyed to a recipient industrial production strain by DNA-mediated transformation and a single chromosome or a few chromosomes can be transferred in *kar1* matings.

The detailed analysis of individual brewers' yeast chromosomes may be illustrated by the extensive studies of Nilsson-Tillgren et al.[11] When meiotic segregants from the Carlsberg lager beer strain were mated to laboratory haploids that carried the *kar1* mutation, as well as other auxotrophic marker genes, it was possible to select progeny that contained a single specific chromosome derived from the wild-type brewing strain. Selections for the extra chromosome required that it contribute a functional wild-type gene complementary to the auxotrophic marker. Subsequent genetic analysis of such "addition lines" entailed further crosses. Occasionally, the *kar1* hybrid loses a cerevisiae chromosome and a substitution line is obtained wherein a chromosome of the tester strain is replaced by a homologous or homeologous chromosome derived from the brewing yeast.

Eight Carlsberg chromosomes homologous or homeologous to cerevisiae chromosomes III, V, XII, and XIII were analyzed. Two different versions of chromosomes V, XII, and XIII were uncovered in the brewing strain. One form was homologous and permitted effectively normal crossing over and molecular hybridization in the tested regions. The homeologous chromosome form carries functionally wild-type alleles as indicated by the complementional response, but molecular hybridization is, at best, weak and meiotic recombination is effectively blocked. Tetrad analysis from crosses involving chromosome III addition or substitution lines revealed that crossing over between the cerevisiae and Carlsberg chromatin was absent in the *his4* to *mat* interval, while recombination in the *thr4-mat* interval was normal. Clearly, significant differences of this sort constitute a rich repository of genetic diversity and this may be utilized advantageously in strain improvement.

Highly diversified, natural polymorphisms available from wild and cultivated yeast strains can readily serve as a natural genetic reservoir for strain improvement. Additionally, these variations may be adapted by the industrial geneticist as a unique taxonomic handle. We refer here to the molecular identification and characterization of yeast strains by recombinant DNA analysis.

Traditionally, industrial strains are characterized according to their morphological and fermentation attributes — parameters that are readily accessible and easily evaluated. Currently, however, strains can be identified more precisely by the use of restriction endonucleases and the identification of unique RFLPs — restriction fragment length polymorphisms (see Kielland-Brandt[12]). Restriction fragments are readily visualized on ethidium bromide-stained electrophoretograms.

Bands corresponding to highly reiterated genomic segments as in the 100-fold repeated rDNA are especially prominent.[13,14] For example, a profile with three larger EcoR1 bands at 3.0, 2.8, and 2.0 kb and three smaller, fainter bands identify pattern II in Carlsberg lager yeast.[14] However, in contrast, pattern I contains an additional EcoR1 site within the 3.0-kb segment. In turn, this generates two smaller fragments of 2.4 and 0.6 kb. A third pattern (III) can be identified in *S. bayanus*, a widely used commercial champagne yeast. Here, the 3.0-kb fragment is shorter by 50 base pairs. Still other patterns will undoubtedly be identified in wild yeasts as surveys become more extensive.

An impressive congery of varied polymorphisms can also be detected by employing probes specific for any of the 1000 or more known yeast genes. This may be illustrated by our own findings centered on the $CUP1^r$ of *S. cerevisiae*.

Recently, we cloned this gene[15] and discovered that the wild-type resistant strain X2180 carried ten tandemly arrayed copies of a 2.0-kb segment at the $CUP1^r$ locus. In contrast, the standard sensitive allele carried only a single copy. The laboratory strain 2.0-kb repeat unit harbors several restriction regions, i.e., single Kpn1 and Xbal sites and two Sau3A sites. DNA sequence studies by Karin et al.[16] revealed that the repeat unit embraced two distinct and different open reading frames. One represents a larger protein of unknown function, and the other a copper binding metallothionein-like protein. The Cu-chelatin is inducible by copper, but not by zinc or cadmium, and overall we could demonstrate that the copper resistance level of the strain is proportional to the gene copy number at the locus. Subsequent studies by Hamer[17] and others concerned largely with the regulation of gene expression in this system confirmed our central findings and also demonstrated that deletion and replacement of the $CUP1^r$ locus did not generate a lethal situation. Hence, the locus does not contribute an essential function to the metabolic economy of the cell. It is, nonetheless, extremely useful. Given these findings, we presumed that wild yeasts and commercial cultivars would reflect rapid and extensive evolutionary changes at the single locus level.

Accordingly, we undertook a genetical and molecular characterization of various industrial yeasts. Included were commercial cultivars representing wine, beer, bread, and distiller's yeast (Table 1). The *Saccharomyces* sp. represented *diastaticus, oviformis, fermentati, bayanus, ludwigii,* as well as some related forms such as *Hansensula wingii* and *Schizosacchromyces pombe*.[18b]

The various yeasts were tested for copper tolerance levels, and to determine whether the phenotypic resistance level was correlated to the genotypic $CUP1^r$ gene copy number, we analyzed genomic DNA samples by Southern blot hybridization using a $CUP1^r$ specific probe. Hybridization patterns for EcoR1-digested genomic DNA obtained from these strains are given in Figure 1. Included for comparison are the resistant laboratory strains X2180 ($CUP1^r$, 14 copies) and $cup1^s$, the single copy-sensitive strain. EcoR1 restriction fragments hybridizing to the specific probe yield a first approximation to the physical extent of the locus, since this enzyme fails to cleave the DNA within the repeat unit. Patently, the various cultivars constitute an impressive array with loci ranging from about 40 to 5 kb or about 17 copies to 1 copy. The findings imply extensive variability with respect to gene copy number. It is noteworthy that some strains carry two or more distinctive loci. For example, strain 522, a Montrachet wine yeast widely used in California commercial red wine production, displays two characteristic EcoR1 bands. One fragment carries about 16 tandem copies, while the other band represents two tandem copies. Fleischman's (FL-1), a commercial baker's yeast, also harbors two bands — the copy numbers are, respectively, four and one.

Table 1
GENE COPY NUMBER AND REPEAT UNIT LENGTH IN VARIOUS YEASTS

Strain		Iterated unit (kb)
513-1	*Saccharomyces cerevisiae* (distillers)	1.6
505-1	*S. bayanus* (champagne)	1.6
519-1	*S. fermentati* (sherry flor)	No repeat
522-1	*S. cerevisiae* (Montrachet)	1.8 & 1.6
Fl-1	*S. cerevisiae* (baker's)	1.8
68-917B	*S. kluyveri*	—
73-164-1	*S. cerevisiae*	1.8
60-322-1	*S. oviformis*	1.6
cc-164-1	*S. diastaticus* (brewer's)	1.1
YNN53	*Saccharomycodes ludwigii*	—
X2180-1A	*Saccharomyces cerevisiae* (lab strain)	2.0

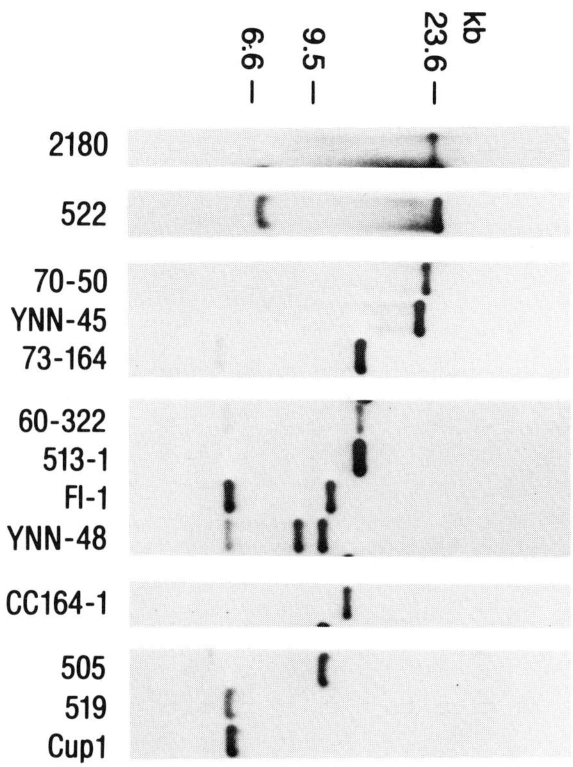

FIGURE 1. DNA hybridizations of EcoR1 digested genomic DNA (20 to 40 μg per lane) from single cell isolates of strains listed. Size markers are given at left.

Patently, the industrial strains contain variously sized loci and in at least five strains the locus is differentially iterated on multiple homologous chromosomes. Also, it may be noted that commercial yeasts are frequently mixtures composed of two or more cell types. The findings reported here, however, are based on cultures derived from single cells isolated by direct micromanipulation. Hence, multiple bands denote chromosomal organization differences. Related to the issue of locus size or extent of iteration are questions impinging on the dimension and form of the repeat unit in our yeast flora sample.

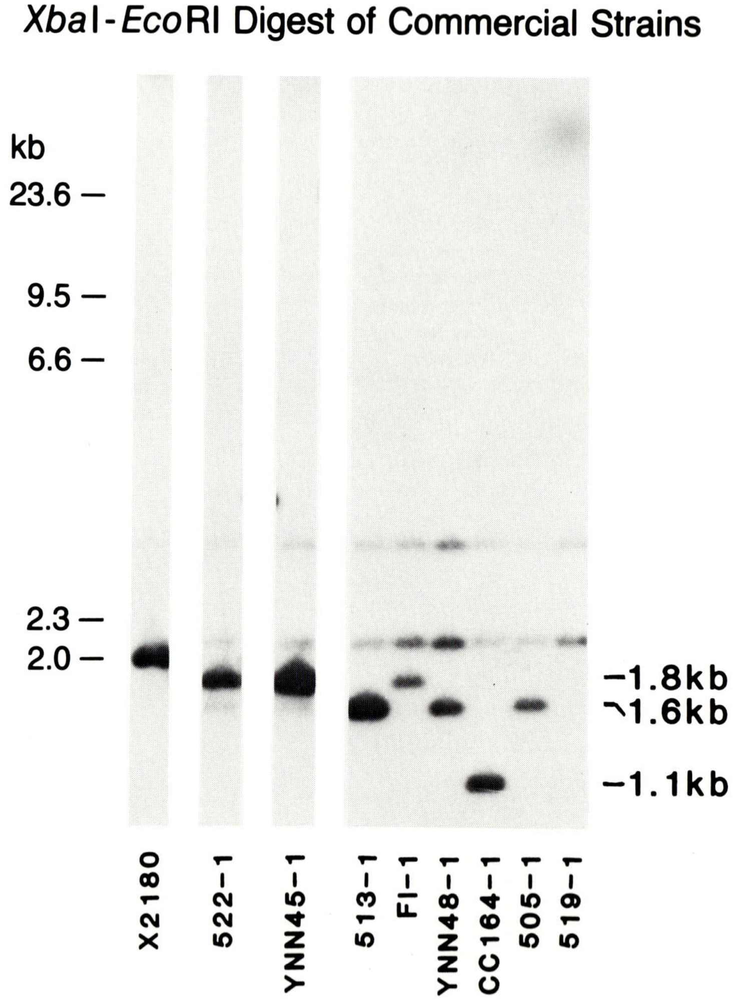

FIGURE 2. DNA hybridizations of XbaI-EcoR1 double digests (20 to 40 μg per lane). All strains have junction lengths of bands of 2.1 and 3.1 kb. Multimolar bands are indicated at right. These correspond to the size of repeat unit(s) in each strain.

Information concerning the repeat unit itself is obtained readily from molecular hybridization studies that utilize doubly digested genomic DNA samples, i.e., cut with two enzymes, EcoR1 and Xbal. Total digestion with these latter endonuclease cleaves the locus bounded by external EcoR1 sites into two constant unimolar flank regions, in addition to a multimolar band representing the iteration unit of which the locus is constituted. Figures 2 and 3 summarize these findings. Taken collectively within our limited sample of eight yeasts, we identified at least three unique repeat units, respectively, 1.8, 1.6, and 1.1 kb in length. Each unit includes the entire coding sequence corresponding to the copper-binding protein, variously designated as copper-chelatin, copper-thionein, or yeast-metallothionein. Only the 1.8- and the 2.0-kb units include repeated copies of the larger unknown gene. The smaller 1.6- and 1.1-kb repeats are variously truncated in this regard. Further strain characterization supplementary to the molecular studies was attempted by sporulation, dissection, and tetrad analysis of the purified isolates. Almost all strains were sporulation competent on KAC median (except YNN48). For each culture, 20 tetrads were dissected and in most instances at least a few viable spores were recovered. The significance attaching to these genetical studies may be illustrated by the observations on two strains.

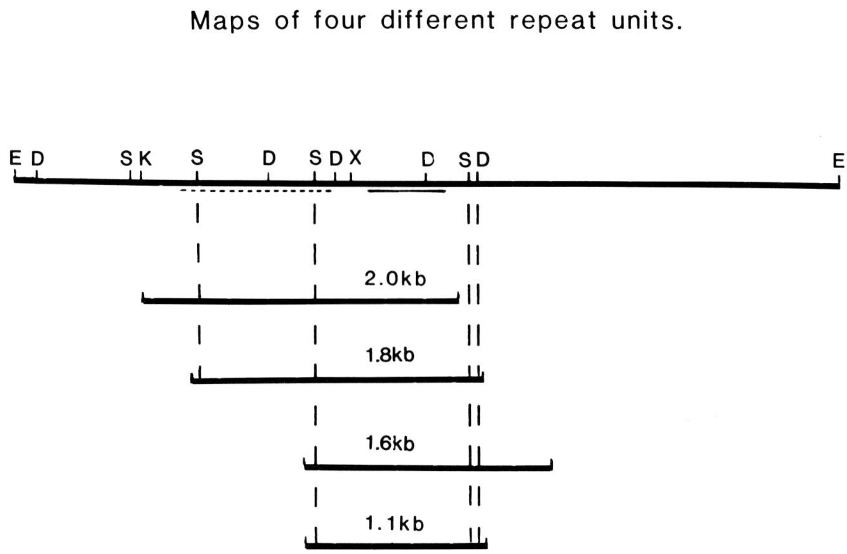

FIGURE 3. Restriction map of the copper locus showing the extent of variously sized repeat units. E = EcoR1, S = Sau3A, K = Kpn1, X = Xba1. The dotted line (---) indicates the position of a transcribable gene of unknown junction. The solid line (—) to the right of the Xba site indicates the position of the copper-chelatin gene. All repeat units contain the entire chelatin gene. Only the 2.0 repeat includes the entire unknown gene. The Xba1-EcoR1 distances remain constant in all strains.

The purified single cell culture from the parental red wine strain 522, sometimes designated as Montrachet, displayed two distinct loci with about 16 and 2 copies, respectively. From single isolates, 6 colonies sporulated after 2 weeks and each of 12 spores representing 3 complete tetrads carried both *CUP1*ʳ loci, respectively, corresponding to the 16 and 2 copies present in the original parent cell. However, these ascosporal colonies were themselves sporulation proficient. Hence, a second tetrad analysis cycle was undertaken and six ascosporal colonies from the secondary sporulation cycle were analyzed further. Each contained either the 16- or the 2-copy locus. Spores carrying both loci were not found. Clearly, the loci segregate in a predictable fashion. Tentatively, we diagnose strain 522 as a homothallic diploid tetrasomic for chromosome VIII. It is particularly noteworthy that the larger 16-copy locus embodies repeat units of size 1.6 and 1.8, while the smaller locus embraces two tandemly arrayed 1.6-units. At present we have no data concerning the relative numbers or spatial distribution of the 1.6- and 1.8-repeat units within the larger locus.

Similar evidence was adduced for the commercial baker's yeast FL-1 (Fleischman's). Genomic DNA from this strain also exhibited two distinctive EcoR1 bands that corresponded to four copies and one copy. The strain sporulated abundantly, though ascospore survival was only about 20%. One complete tetrad was recovered and subjected to further analysis. Two spore colonies carried only the single copy locus and the two remaining spore derivatives carried both loci, but were sporulation deficient. Thus, preliminary tetrad data support the view that differentially amplified *CUP1*ʳ loci may reside on homologous chromosome, and these display normal Mendelian segregation patterns. Finally, that all *CUP1*ʳ gene copies are localized distal to *thr1* on chromosome VIII is supported by the following observation. In all strains tested, the restriction pattern in the 10 to 15 kb of flanking DNA is essentially the same, except for an occasional single site polymorphism. Though our analysis is partial and incomplete, on a minimal hypothesis, we assume that this baker's yeast is probably diploid and likely trisomic for chromosome VIII. Additional aspects concerning the extent of molecular and genetic variability in brewer's yeast was reviewed *in extenso* by von Wettstein et al.,[4] Tubb,[19] and Stewart and Russell.[20]

The first successful DNA-mediated transformation in yeast was reported by Hinnen et al. in 1978.[21] Yeast protoplasts were exposed to a plasmid containing foreign DNA (carried as pYe leu2) in the presence of PEG. The recipient spheroplasted yeast strain carried a double mutant at *leu2* and *LEU*$^+$ prototrophs were selected by platings on appropriate selective media. Three categories were identifiable among the recovered transformants as follows: integrations of the plasmid into a chromosome III sequence complementary to *LEU2*$^+$, plasmid integrations into other unlinked chromosomal sites, and complete transfers of *LEU2*$^+$ information unaccompanied by bacterial DNA into the corresponding *leu2*$^-$ region. In effect, the latter event is a gene conversion wherein the double mutant chromosomal sequence is replaced by the corresponding plasmid wild-type sequence at *LEU2*.

At about the same time the 2u plasmid derived from yeast was incorporated by Beggs[22] into a chimeric plasmid pJDB41. Thus, the 2u plasmid and pMB9 which carried a selectable tetracycline-resistance marker sequence were combined. This derivative hybrid plasmid transformed *Escherichia coli* cells from which the highly amplified plasmid was subsequently isolated. Cut with PstI and tailed with poly dA, the plasmid DNA was ligated with sheared, linear, wild-type yeast DNA carrying similar tails. Thus, *leu*$^-$ bacteria could be transformed to their *Leu*$^+$ counterparts. A particular plasmid carried a 1.2-kb insertion and this chimeric plasmid could regularly transform yeast *leu2* spheroplasts to wild type at high frequencies — about 1/1000[3] to 1/10,000 per viable cell.

The literature concerning transformation and recombinant DNA strategies in yeast is extensive and highly varied. It impinges on virtually every aspect of molecular genetical science. Boguslawski's review[23] provides an excellent introduction to the organized body of research concerned with general principles, methods, vectors, artificial chromosomes, gene isolation, and characterization as well as the expression of heterologous genes in yeast. The power of these approaches is highly sophisticated, though still at a developmental phase close to infancy.

Now, we may briefly consider the fact that numerous useful genes, some of mammalian origin and otherwise coding for enzymes, hormones, and various pharmaceuticals, are approaching commercial production levels in yeast. Insulin somatostatin, epidermal growth factor, rennin uricase, interferon, hepatitis vaccine, and perhaps two dozen or more products will be marketed in the immediately foreseeable future, a prediction based on the gleanings from numerous annual reports and Wall Street's prognosticator savants. Table 2, taken from Herman Phaff's (1981) *Scientific American* article on industrial microorganisms, is especially prophetic both with respect to the array of organisms that have been harnessed and the range of products. The gene pool in industrial and wild yeasts is wide, deep, and virtually unplumbed. To this we must add the gene pools of 500 other yeast species. Furthermore, that yeast is an organism uniquely adapted to the needs of biotechnology arises from a constellation of happy facts. As a production vehicle it does not generate bacterial products; also, it glycosylates proteins in a manner similar to the mammalian glycosylation pattern, and it possesses a precisely controllable protein secretion mechanism that facilitates the recovery of genetically engineered end products at the industrial-scale level. Capital-intensive developments in this field appear as commonplace feature articles or new items on the daily financial pages. More to the point, rising on the landscape of today are new yeast fermentors with capacities measured in tens of thousands of gallons. The first major transitional phase has ended. Molecular geneticists and entrepreneurs, having recombined and replicated, are no longer circling in the dark as they split their stocks under the watchful eye of the Security Commission.

In closing, it is fitting and appropriate to recall a bit of relevant history. The first reported heritable transformation of *Diplococcus pneumoncae* (redesignated as a *Streptococcus*) with heat-killed strains possessing a diagnostically distinguishable extracellular polysaccharide occurred in 1928. Some 16 years elapsed before it was convincingly demonstrated that DNA, and DNA effectively without contaminating protein, could effect the transformation reaction.

Table 2
PRODUCTS FROM VARIOUS TYPES OF INDUSTRIAL MICROORGANISMS

Organism	Type	Product
	Foods and Beverages	
Saccharomyces cerevisiae	Yeast	Baker's yeast, wine, ale, sake
S. carlsbergensis	Yeast	Lager beer
S. rouxii	Yeast	Soy sauce
Candida milleri	Yeast	Sour French bread
Lactobacillus sanfrancisco	Bacterium	Sour French bread
Streptococcus thermophilus	Bacterium	Yogurt
L. bulgaricus	Bacterium	Yogurt
Propionibacterium shermanii	Bacterium	Swiss cheese
Gluconobacter suboxidans	Bacterium	Vinegar
Penicillium roquefortii	Mold	Blue-veined cheeses
P. camembertii	Mold	Camembert and brie cheeses
Aspergillus oryzae	Mold	Sake (rice-starch hydrolysis)
Rhizopus	Mold	Tempeh
Mucor	Mold	Sufu (soybean curd)
Monascus purpurea	Mold	Ang-kak (red rice)
	Industrial Chemicals	
Saccharomyces cerevisiae	Yeast	Ethanol (from glucose)
Kluyveromyces fragilis	Yeast	Ethanol (from lactose)
Clostridium acetobutylicum	Bacterium	Acetone and butanol
A. niger	Mold	Citric acid
Xanthomonas campestris	Bacterium	Polysaccharides
	Amino Acids and Flavor-Enhancing Nucleotides	
Corynebacterium glutamicum	Bacterium	L-Lysine
C. glutamicum	Bacterium	5′-Inosinic acid and 5-guanylic acid
	Single-Cell Proteins	
Candida utilis	Yeast	Microbial protein from paper-pulp waste
Saccharomycopsis lipolytica	Yeast	Microbial protein from petroleum alkanes
Methylophilus methylotrophus	Bacterium	Microbial protein from growth on methane or methanol
	Vitamins	
Eremothecium ashbyi	Yeast	Riboflavin
Pseudomonas denitrificans	Bacterium	Vitamin B_{12}
Propionibacterium	Bacterium	Vitamin B_{12}
	Enzymes	
A. oryzae	Mold	Amylases
A. niger	Mold	Glucamylase
Trichoderma reesii	Mold	Cellulase
Saccharomyces cerevisiae	Yeast	Invertase
K. fragilis	Yeast	Lactase
Saccharomycopsis lipolytica	Yeast	Lipase
Aspergillus	Mold	Pectinases and proteases
Bacillus	Bacterium	Proteases
Endothia parasitica	Mold	Microbial rennet

Table 2 (continued)
PRODUCTS FROM VARIOUS TYPES OF INDUSTRIAL MICROORGANISMS

Organism	Type	Product
	Polysaccharides	
Leuconostoc mesenteroides	Bacterium	Dextran
X. campestris	Bacterium	Xanthan gum
	Pharmaceuticals	
Penicillium chrysogenum	Mold	Penicillins
Cephalosporium acremonium	Mold	Cephalosporins
Streptomyces	Bacterium	Amphotericin B, kanamycins, neomycins, streptomycin, tetracyclines, and others
B. brevis	Bacterium	Gramicidin S
B. subtilis	Bacterium	Bacitracin
B. polymyxa	Bacterium	Polymyxin B
Rhizopus nigricans	Mold	Steroid transformation
Arthrobacter simplex	Bacterium	Steroid transformation
Mycobacterium	Bacterium	Steroid transformation
Hybridomas	—	Immunoglobulins and monoclonal antibodies
Mammalian cell lines	—	Interferon
Escherichia coli (via recombinant-DNA technology)	Bacterium	Insulin, human growth hormone, somatostatin, interferon
	Carotenoids	
Blakeslea trispora	Mold	β-Carotene
Phaffia rhodozyma	Yeast	Astaxanthin
	Entomopathogenic Bacteria	
Bacillus thuringiensis	Bacterium	Bioinsecticides
B. popilliae	Bacterium	Bioinsecticides

From Phaff, H., *Sci. Am.*, 1981.

Despite exacting proofs, 10 years later geneticists were still pondering whether the particle of genetics could be protein or, perhaps, a protein-DNA complex. Since DNA was then assumed to be rather simple, i.e., a tetranucleotide in character, it could not encompass the genetic variability known to exist within a single gene. Thus, cherished though unwarranted assumptions with keen disappointments were met and science moved forward to the present chapter. The concluding verse of Rollin D. Hotchkiss's poem "(R)Evolution" is a fitting preachment:[24]

> Now, least we be too placid—
> There might be other genes
> That instead of nucleic acid
> Are made up of proteins
> How perverse our limitations — how incredibly tragic
> That, for each everyone of us one or the other of these two
> Hypothesis should seem relatively acceptable and the
> Alternative one should seem like believing in magic.

REFERENCES

1. **Spencer, J. F. T.,** Genetic improvements of industrial yeasts, *Ann. Rev. Microbiol.,* 37, 121, 1983.
2. **Stewart, G. G.,** The genetic manipulation of industrial yeast strains, *Can. J. Microbiol.,* 27, 973, 1981.
3. **Johnston, J. R. and Oberman, H.,** Yeast genetics in brewing, in *Progress in Industrial Microbiology,* Vol. 15, Bull, M. J., Ed., Elsevier, Amsterdam, 1979, 151.
4. **von Wettstein, D., Gjermansen, C., Holmberg, S., Kielland-Brandt, M. C., Nilsson-Tillgren, T., Pedersen, M. B., Petersen, J. G. L., and Sigsgaard, P.,** Genetic Engineering in the Improvement of Brewer's Yeast, an address presented at the MBAA convention, San Diego, Calif., October 23 to 26, 1983.
5. **Anderson, E. and Martin, P. A.,** Sporulation and mating of brewer's yeasts, *J. Inst. Brew.,* 81, 242, 1983.
6. **Reader, H. P.,** Genetic Analysis of Flocculation is *Saccharomyces cerevisiae,* Ph.D. thesis, University of Strathclyde, Glasgow, 1980.
7. **Spencer, J. F. T., Laud, P., and Spencer, D. M.,** The use of mitochondrial mutants in the isolation of hybrids involving industrial yeast strains. II. Use in isolating hybrids obtained by protoplast fusion, *Mol. Gen. Genet.,* 178, 651, 1981.
8. **Ferenczy, L.,** Microbial protoplast fusion, *Symp. Soc. Gen. Microbiol.,* 31, 1, 1981.
9. **Weber, H., Furster, W., Jacob, H. E., and Berg, H.,** *Z. Allg. Mikrobiol.,* 21, 555, 1980.
10. **Conde, J. and Fink, G. R.,** A mutant of *Saccharomyces cerevisiae* defective for nuclear fusion, *Proc. Natl. Acad. Sci. U.S.A.,* 79, 5342, 1982.
11. **Nilsson-Tillgren, T., Gjermansen, C., Kielland-Brandt, M. C., Petersen, J. G. L., and Holmberg, S.,** Genetic differences between *Saccharomyces carlsbergenesis* and *S. cerevisiae*. Analysis of chromosome III by single chromosome transfer, *Carlsberg Res. Commun.,* 46, 65, 1981.
12. **Kielland-Brandt, M. C.,** The genetics of brewer's yeast, *Proc. Eur. Brew. Conv. (Copenhagen),* p. 236, 1981.
13. **Carlson, M. and Botstein, D.,** Organization of the *SUC* gene family in *Saccharomyces, Mol. Cell. Biol.,* 3, 351, 1983.
14. **Petes, T. D. and Botstein, D.,** Simple mendelian inheritance of the reiterated ribosomal DNA of yeast, *Proc. Natl. Acad. Sci. U.S.A.,* 74, 5091, 1977.
15. **Fogel, S. and Welch, J. W.,** Tandem gene amplification mediates copper resistance in yeast, *Proc. Natl. Acad. Sci. U.S.A.,* 79, 5342, 1982.
16. **Karin, M., Najarian, R., Haslinger, A., Valenzuela, P., Welch, J. W., and Fogel, S.,** Primary structure and transcription of an amplified genetic locus: the *CUP1* locus of yeast, *Proc. Natl. Acad. Sci. U.S.A.,* 81, 337, 1984.
17. **Hamer, D. H., Thiele, D. J., and Lemontt, J. E.,** Function and autoregulation of yeast copperthionein, *Science,* 228, 685, 1985.

18a. **Welch, J. W., Fogel, S., Cathala, G., and Karin, M.,** Industrial yeasts display tandem gene iteration at the *CUP1* region, *Mol. Cell. Biol.,* 3, 1353, 1983.

18b. **Fogel, S. and Welch, J. W.,** A recombinant DNA strategy for characterizing industrial yeast strains, *Proc. 15th Int. Congr. Genet. New Delhi,* p. 133, 1983.
19. **Tubb, R. S.,** Applying yeast genetics. A current assessment, *J. Inst. Brew.,* 85, 286, 1979.
20. **Stewart, G. G. and Russell, I.,** Current use of the "new" genetics in research and development of brewer's yeast strains, *Eur. Brew. Conv. Congr. Berlin,* p. 475, 1979.
21. **Hinnen, A., Hicks, J., and Fink, G. R.,** Transformation in yeast, *Proc. Natl. Acad. Sci. U.S.A.,* 75, 1929, 1978.
22. **Beggs, J. D.,** *Nature,* 275, 104, 1978.
23. **Boguslawski, G.,** *Yeast Transformation in Gene Manipulations in Fungi,* Academic Press, New York, 1985.
24. **Fristrom, J. and Spieth, P., Eds.,** *Principles of Genetics,* Chiron Press, New York, 1980, 3.

Chapter 8

BREEDING FOR ETHANOL-TOLERANT DISTILLER'S YEAST

Bjorn Eggert Christensen

TABLE OF CONTENTS

I. INTRODUCTION

In the distilling industry there is an obvious desire for yeasts that have a high fermentation rate, high ethanol tolerance, and high viability. Hybridization and selection are important tools for the improvement of other yeast strains, such as those used in the brewing,[11,17,30] baking,[12,31,32] and winemaking[9,26,29] industries.

Mutagenesis has been used in the breeding of ethanol-tolerant wine yeasts[1] and selective hybridization has produced hybrids with a better wine-producing potential.[9,29] Ethanol-resistant variants have been recovered by selection in continuous cultures[4,19] and protoplast fusion has been used to create strains with improved fermentation characteristics.[16,25]

The response of yeast cells to ethanol has been shown to be strain dependent and affected by many genes.[6,9,14,15,29] In addition to the genome, the environment is important for ethanol tolerance. Osmotic pressure,[23] the presence of unsaturated fatty acids and sterols,[3,5,27,28] and temperature[13,20,21] influence the effect of ethanol on yeast cells.

So far, sophisticated breeding has been hampered by insufficient knowledge of the genetics of industrial yeast strains. Aneuploidy and/or polyploidy have been observed in several strains[8,10,12] and structural heterozygosity, leading to reduced recombination and nondisjunction, may be a reason for low spore viability in brewer's yeast.[22] Reduced spore viability, coupled with a majority of the germinating spores being nonmaters, makes traditional genetic analysis difficult. Keiding[18] has shown by tetrad analysis of restriction endonuclease fragment polymorphisms that spore clones of the *Saccharomyces cerevisiae* strain DDSF 102, used in this study, were disomic for at least seven chromosomes, and that four different copies of chromosome III are present in the strain. Of the four chromosomes III, three were lacking HML and one also lacked HMR. These results indicate that DDSF 102 is tetraploid and heterozygous for a number of tested chromosome regions. The tested loci of strain DDSF 102 were with regard to their nucleotide sequence highly homologous to those of genetically characterized standard strains.[18]

The acquired detailed genetic knowledge about strain DDSF 102 as well as the finding that it contains HO, a gene for homothallism, can now be taken into account in efforts to improve the performance of this strain by using its inherent genetic variation. By random mating of a large number of spores many new combinations will arise. Cells with desirable properties can then be isolated if a suitable selection procedure can be deployed.

A breeding program for ethanol-tolerant yeasts is carried out by De Danske Spritfabrikker: both pedigree breeding and random breeding by mass mating followed by selection[6,7] are employed.

II. SELECTION AMONG PROGENY OF A SINGLE STRAIN

Selection among sexually produced progeny of *Saccharomyces uvarum* strains for improved biomass production has been reported by Azevedo et al.[2] They adapted methods currently used in breeding of higher plants, according to the principles of quantitative genetics. Their results indicate that the use of the existing natural variation can lead to significant genetic improvements. Crossbreeding of brewer's yeast[11] leads to the same conclusion.

A similar approach has been tried in the breeding for enhanced ethanol tolerance. Ascospores from the production strain DDSF 102 were mixed on a nutrient agar plate and allowed to germinate, thus facilitating mass mating of spore clones. The mated mixture was propagated in diluted beet molasses, harvested, and transferred to a fermentation flask for the selection for increased viability after fermentation in molasses.[7]

The progress of the selection is shown in Figure 1. The degree of fermentation is calculated from the recorded weight losses of the fermentation flasks as the percentage of sucrose

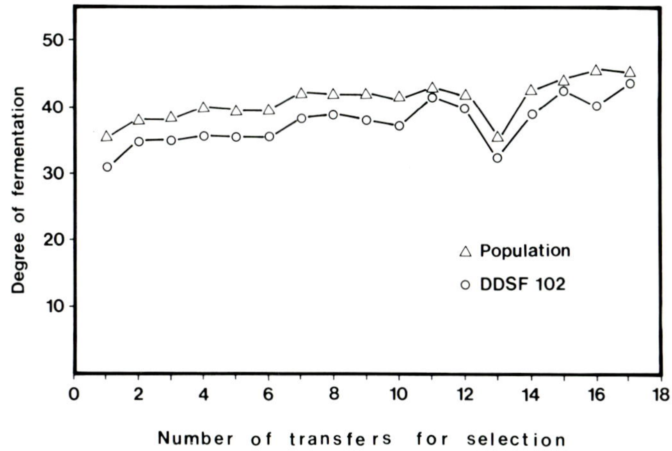

FIGURE 1. Progress of the degree of fermentation during selection for a population derived from meiotic progeny of strain DDSF 102 by random mass mating. The degree of fermentation was calculated from the weight loss of the fermentation flasks as the percentage of sucrose fermented. Yeast cells from the first selection experiment were used as inoculum in the second etc. The marked drop in fermentation ability at experiment #13 coincides with the use of a new batch of medium. The medium was 37.5% (w/v) beet molasses, 12% (w/v) sucrose, pH 5.0, and the temperature 37°C.

fermented. Both the control, strain DDSF 102, and the population undergoing selection achieve a progressively higher degree of fermentation; strain DDSF 102 probably by adaptation alone and the population by both adaptation and selection for better ethanol producers. The relative low values for the attenuation of the molasses wort are explained by the conditions in the medium: high osmotic pressure, high temperature, and the presence of substances in the molasses that are known to be inhibitory to yeast. Still almost half of the sucrose is converted during the experiment, giving a final ethanol concentration of approximately 8% (v/v).

Cell material from the final selection flask was plated on molasses plates and 64 vigorously growing colonies were collected. These isolated clones were tested individually for fermentation performance at 37°C in small-scale batch fermentations with molasses as substrate.[7]

After 20 hr the mean value of the degree of fermentation had shifted towards a higher value (Figure 2). The same applies to the values after 44 hr. A substantial part of the selected strains possesses fermentation abilities considerably exceeding those of strain DDSF 102.

Table 1 gives the degrees of fermentation at both 20 and 44 hr for the best of the isolated clones. The final degree of fermentation has been improved by 12% on the average, reflecting a rise in final ethanol concentration from 10 to 11.6% (v/v).

Heterozygosity is still present in these selected strains.[7] Strain M2/5-23, equal in performance to M2/5-32, was sporulated and the spore progenies again subjected to the selection procedure; a few of the isolated 60 clones expressed fermentation traits significantly better than those of the parent strain M2/5-23.[7]

Korhola[19] and Brown and Oliver[4] obtained ethanol tolerant yeasts by selection in continuous cultures. Stable ethanol-tolerant strains were produced by the addition of ethanol in a chemostat. The resulting strains gave higher rates of fermentation and ethanol concentrations than their parent strains. In both studies, defined media using glucose as carbon source were employed. In a study parallel to the present, yeast strains selected on a defined sucrose medium performed poorly on molasses media. None of the strains, which in the defined medium exceeded DDSF 102, gave ethanol concentrations in the same range as the parent

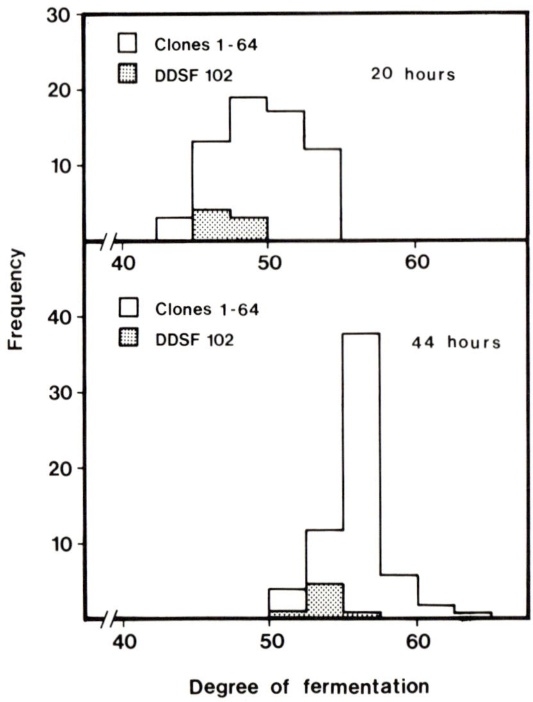

FIGURE 2. Frequency distributions of the fermentation abilities of 64 selected clones, determined in small-scale batch fermentations. The degree of fermentation is calculated from weight losses after 20 and 44 hr. The medium was 37.5% (w/v) beet molasses, 12% (w/v) sucrose, pH 5.0, and the temperature 37°C.

strain when tested on molasses media, indicating the necessity of direct selection in the medium to be used in the production process.

III. EFFICIENCY IN ETHANOL PRODUCTION

A number of the strains presented in Table 1 were further evaluated by continuous fermentation in a bench top fermenter. The yeast was propagated for 24 hr under aerobic conditions in the fermenter, followed by continuous fermentation on molasses media for 7 days. Figure 3 shows the typical progress of fermentation for the selected strain M2/5-32 and the control, DDSF 102. The ethanol concentration and viability remain at a higher level for the selected clone than for DDSF 102. This may be due to a higher productivity or to a higher viability.

Strain M2/5-32 was tested in a full-scale fermentation. The production equipment consisted of three tanks of 180 m^3 each, connected in series. The fermentation was continuous and yeast harvested from the fermented wort was recycled.[24] Figure 4 gives the ethanol concentrations during the fermentation in the three tanks. Strain DDSF 102 gives the lowest ethanol concentration in all tanks, i.e., 5 to 6% (v/v) in tank 1, 7 to 8% (v/v) in tank 2, reaching the highest level, 8.7% (v/v), in tank 3. Noteworthy is the tendency of decreasing ethanol production during the second half of the fermentation. Strain M2/5-32 attained ethanol concentrations at least 1% higher than those obtained with DDSF 102. The concentration in tank 1 reached 7% and the final ethanol concentration was about 10% (v/v). The small difference between the ethanol concentrations in tank 2 and 3 indicates that ethanol formation

Table 1
FERMENTATION PARAMETERS FOR DDSF 102 AND 22 CLONES ISOLATED FROM A SELECTED POPULATION

Strain	Degree of fermentation (20 hr, mean)	Degree of fermentation (44 hr, mean)
DDSF 102	46.8	53.1
M2/5-1	53.6	59.8
M2/5-4	52.6	56.7
M2/5-5	52.2	63.1
M2/5-8	52.9	55.8
M2/5-12	52.7	55.9
M2/5-15	52.0	56.5
M2/5-20	52.2	60.2
M2/5-23	52.1	56.5
M2/5-24	52.5	56.4
M2/5-32	53.6	57.5
M2/5-33	49.2	56.9
M2/5-34	52.7	54.9
M2/5-36	53.0	55.3
M2/5-37	51.9	59.8
M2/5-38	51.9	56.2
M2/5-39	53.6	56.9
M2/5-42	52.8	55.1
M2/5-46	50.2	56.2
M2/5-47	51.3	55.2
M2/5-51	52.5	56.2
M2/5-52	52.6	55.0
M2/5-59	52.3	54.9

Note: The fermentation tests were carried out as small-scale batch fermentations. The degree of fermentation was determined from the weight losses of the fermentation flasks as the percentage of sucrose fermented. The medium was 37.5% (w/v) beet molasses supplemented with 12% (w/v) sucrose, pH 5.0; fermentation temperature was 37°C.

is almost complete in tank 2 and that the yeast has a capacity to convert more molasses to higher concentrations of ethanol.

In a previous full-scale experiment using molasses with a higher sucrose content than that used in this experiment, ethanol concentrations of 11% (v/v) were obtained with M2/5-32, whereas DDSF 102 did not exceed 8.5% (v/v).

The increase in yeast biomass during fermentation, Figure 5, indicates an equal rate of growth for the two strains. This is in marked contrast to the viability data presented in Figure 6. The number of live M2/5-32 cells is only affected by the fermentation to a lesser extent, whereas those of DDSF 102 decrease rapidly. M2/5-32 is thus well suited for prolonged and repeated fermentations. The ethanol output for the two yeasts is shown in Table 2. The difference in yield is not significant, indicating that M2/5-32 did not retain its high viability by producing biomass and sacrifying ethanol yield.

IV. CONCLUSION

The yeast strain DDSF 102 is tetrasomic for many, if not all, chromosomes and heterozygous for several tested chromosome regions.[18] Therefore, random mating of a large number of spores will yield many new combinations of genes, some of which could allow expression of improved fermentation characteristics.

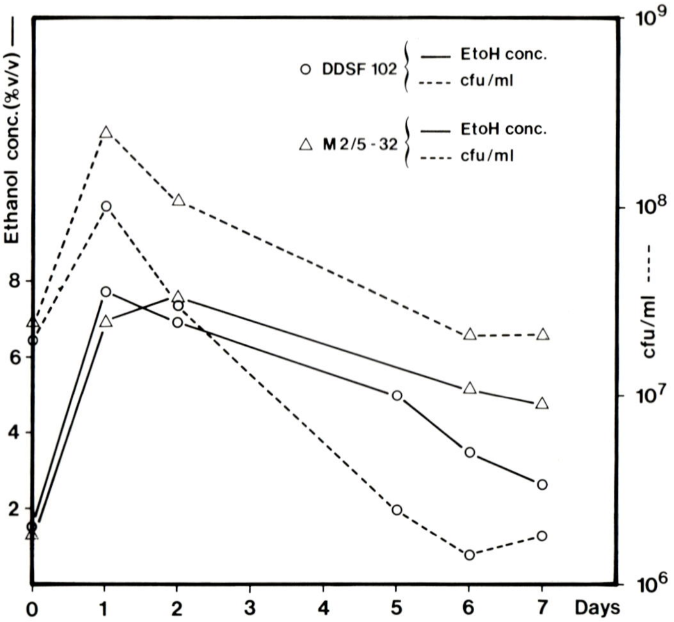

FIGURE 3. Fermentation and viability for DDSF 102 and M2/5-32 in continuous culture. Medium was 30% (w/v) beet molasses, pH 5.0, and the fermentation temperature 32°C. Fermenter volume was 1400 mℓ, rate of feeding 70 mℓ-h^{-1}. cfu = colony forming units.

Strains possessing high tolerance of ethanol, improved fermentation ability, and enhanced viability have been isolated. The present results reveal that selection on a randomly bred population yields yeast strains with the improved characteristics desired by the distilling industry.

ACKNOWLEDGMENTS

The encouragement by and helpful discussions with Diter von Wettstein, Knut Rosen, Morten Kielland-Brandt, and Bjarne Adamsen are greatly appreciated. I wish to thank Kaj Jensen for running the bench-top fermenter experiments, Henrik Eriksen for managing the full-scale trial, and Anne Sahlgren for outstanding technical assistance.

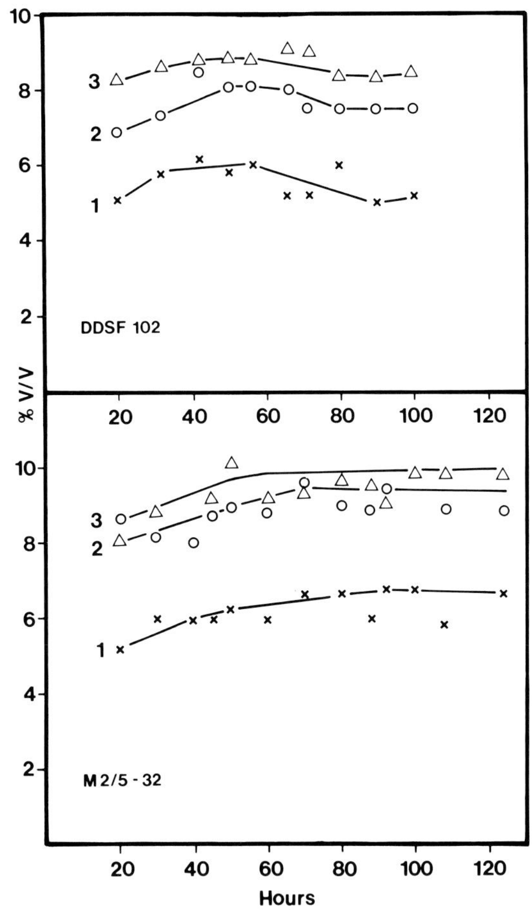

FIGURE 4. Ethanol concentration in tanks 1, 2, and 3 during fermentation with DDSF 102 and M2/5-32. Residual sugar concentration in tank 3 was less than 0.3% (w/v).

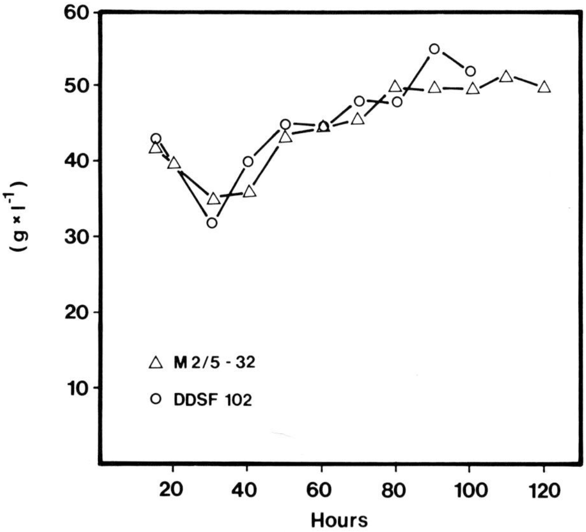

FIGURE 5. Yeast biomass concentration in tank 3 during fermentation for DDSF 102 and M2/5-32. The marked drop at 30 hr indicates the start of yeast recycling.

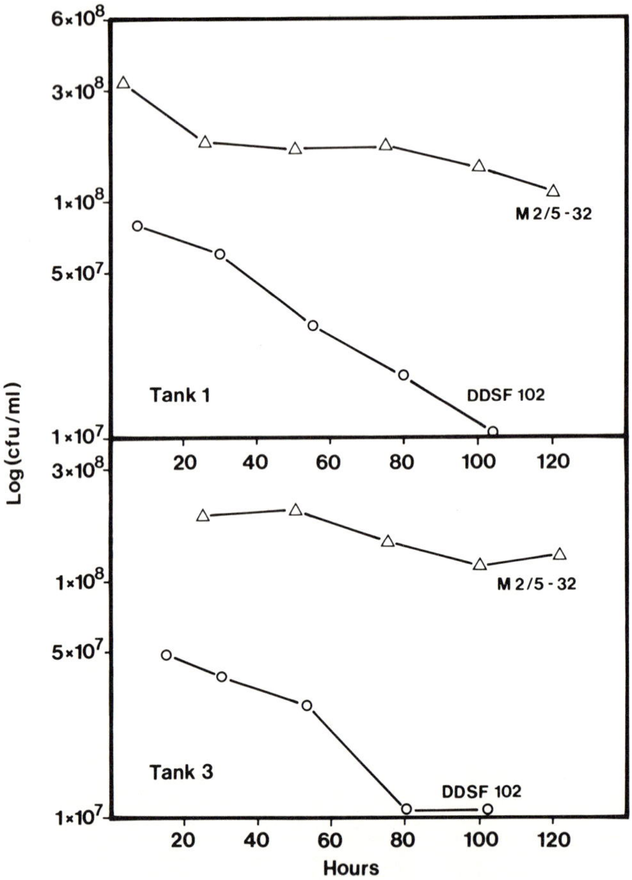

FIGURE 6. Viability of yeast strains DDSF 102 and M2/5-32 during fermentation. Tank 1 was in both experiments inoculated with the same amount of yeast (30 g/ℓ). The number of live cells are scored as colony forming units (cfu): microscopy ensured that the distribution of cells per cfu was equal for the two strains.

Table 2
ETHANOL YIELDS FOR STRAINS DDSF 102 AND M2/5-32. THE YIELDS ARE COMPUTED FROM THE AMOUNT OF MOLASSES USED AND THE AMOUNT OF ETHANOL PRODUCED (MEASURED AFTER THE DESTILLATION)

	DDSF 102	M2/5-32
Molasses used (ton)	705	920
Fermentable sugar in molasses (%)	46.3	44.9
Ethanol produced (1000 ℓ)	195	254
Ethanol yield, ℓ/ton molasses	277	276
Ethanol yield, ℓ/ton fermentable sugar	598	615

REFERENCES

1. **Alikhanyan, S. I. and Nalbandyan, G. M.,** Selection of wine yeasts using mutagens, *Sov. Genet.*, 2, 1200, 1971.
2. **Azevedo, J. L., Tavares, F. C. A., and Melo Cruz, M. R.,** Genetic improvements of yeasts, in *Biochemistry and Genetics of Yeasts,* Bacila, M., Horecker, B. L., and Stoppani, A. O. M., Eds., Academic Press, New York, 1978, 563.
3. **Beaven, M. J., Charpentier, C., and Rose, A. H.,** Production and tolerance of ethanol in relation to phospholipid fatty-acyl composition in *Saccharomyces cerevisiae* NCYC 431, *J. Gen. Microbiol.*, 128, 1447, 1982.
4. **Brown, S. W. and Oliver, S. G.,** Isolation of ethanol tolerant mutants of yeast by continuous selection, *Eur. J. Microbiol. Biotechnol.*, 16, 119, 1982.
5. **Casey, G. P., Magnus, C. A., and Ingledew, W. M.,** High-gravity brewing: effects of nutrition on yeast composition, fermentative ability and alcohol production, *Appl. Environ. Microbiol.*, 48, 639, 1984.
6. **Christensen, B. E.,** Breeding of distillers yeast by hybridization of spore derived clones, *Carlsberg Res. Commun.*, in preparation.
7. **Christensen, B. E.,** Selection for ethanol tolerant distillers yeasts, *Carlsberg Res. Commun.*, in preparation.
8. **Emeis, C. C.,** Polyploide Kulturhefen, *Eur. Brew. Conv. Congr. Wien,* p. 205, 1961.
9. **Eschenbruch, R., Cresswell, K. J., Fischer, B. M., and Thornton, R. J.,** Selective hybridisation of pure culture wine yeasts. I. Elimination of undesirable wine-making properties, *Eur. J. Appl. Microbiol. Biotechnol.*, 14, 155, 1982.
10. **Fowell, R. R.,** Sporulation and hybridization of yeasts, in *The Yeasts,* Vol. 1, Rose, A. H. and Harrison, J., Eds., Academic Press, New York, 1969, 303.
11. **Gjermansen, C. and Sigsgaard, P.,** Construction of a hybrid brewing yeast, *Carlsberg Res. Commun.*, 46, 1, 1981.
12. **Gunge, N.,** Breeding of bakers yeast — determination of the ploidy and an attempt to improve practical properties, *Jpn. J. Genet.*, 41, 203, 1966.
13. **Hacking, A. J., Taylor, I. W. F., and Hanas, C. M.,** Selection of yeast able to produce ethanol from glucose at 40°C, *Appl. Microbiol. Biotechnol.*, 19, 361, 1984.
14. **Ismail, A. A. and Ali, A. A. M.,** Selection of high ethanol yielding *Saccharomyces.* I. Ethanol tolerance and the effect of training in *Saccharomyces cerevisiae, Folia Microbiol.*, 16, 346, 1971.
15. **Ismail, A. A. and Ali, A. A. M.,** Selection of high ethanol yielding *Saccharomyces.* II. Genetics of ethanol tolerance, *Folia Microbiol.*, 16, 350, 1971.
16. **Johansson, M. and Sjöström, J.-E.,** Ethanol and glycerol production under aerobic conditions by wild-type, respiratory deficient mutants and a fusion product of *Saccharomyces cerevisiae, Appl. Microbiol. Biotechnol.*, 20, 105, 1984.
17. **Johnston, J. R.,** Breeding yeasts for brewing. II. Production of hybrid strains, *J. Inst. Brew.*, 71, 135, 1965.
18. **Keiding, A. K.,** Genetic and molecular characterization of a distillers yeast, *Carlsberg. Res. Commun.*, 50, 95, 1985.
19. **Korhola, M.,** Improvement of yeast strains for added ethanol tolerance, in *Gene Expression in Yeast. Proc. of the Alco Yeast Symp.*, Korhola, M. and Vaisänen, E., Eds., Helsinki, 1983, 231.
20. **Nagodawithana, T. W., Castellano, C., and Steinkraus, K. H.,** Effect of dissolved oxygen, temperature, initial cell count and sugar concentration on the viability of *Saccharomyces cerevisiae* in rapid fermentations, *Appl. Microbiol.*, 28, 383, 1974.
21. **Nagodawithana, T. W. and Steinkraus, K. H.,** Influence of the rate of ethanol production and accumulation on the viability of *Saccharomyces cerevisiae* in rapid fermentation, *Appl. Environ. Microbiol.*, 31, 158, 1976.
22. **Nilsson-Tillgren, T., Gjermansen, C., Kielland-Brandt, M. C., Petersen, J. G. L., and Holmberg, S.,** Genetic differences between *Saccharomyces carlsbergensis* and *S. cerevisiae.* Analysis of chromosome III by single chromosome transfer, *Carlsberg. Res. Commun.*, 46, 65, 1981.
23. **Panchall, G. J. and Stewart, G. G.,** The effect of osmotic pressure on the production and excretion of ethanol and glycerol by a brewing yeast strain, *J. Inst. Brew.*, 86, 207, 1980.
24. **Rosen, K.,** Continuous production of ethanol, *Process. Biochem.*, 25, 1978.
25. **Seki, T., Myoga, S., Limtong, S., Vedono, S., Kumnuanta, J., and Taguchi, J.,** Genetic construction of yeast strains for high ethanol production, *Biotechnol. Lett.*, 5, 351, 1983.
26. **Snow, R.,** Toward genetic improvement of wine yeast, *Am. J. Enol. Vitic.*, 30, 33, 1979.
27. **Thomas, D. S., Hossack, J. A., and Rose, A. H.,** Plasma membrane lipid composition and ethanol tolerance in *Saccharomyces cerevisiae, Arch. Microbiol.*, 117, 239, 1978.
28. **Thomas, D. S. and Rose, A. H.,** Inhibitory effect of ethanol on growth and solute accumulation by *Saccharomyces cerevisiae* as affected by plasma-membrane lipid composition, *Arch. Microbiol.*, 22, 49, 1979.

29. **Thornton, R. J.,** Selective hybridisation of pure culture wine yeasts. II. Improvement of fermentation efficiency and inheritance of SO_2 tolerance, *Eur. J. Appl. Microbiol. Biotechnol.*, 14, 159, 1982.
30. **von Wettstein, D., Gjermansen, C., Holmberg, S., Kielland-Brandt, M. C., Nilsson-Tillgren, T., Pedersen, M. B., Petersen, J. G. L., and Sigsgaard, P.** Genetic engineering in the improvement of brewers yeast, *MBBA Tech. Q.*, 21, 45, 1984.
31. **Windisch, S.,** Erfahrungen auf dem Wege, neue Hefestämme für technishe Zwecke zu finden und zu züchten, *Branntweinwirtschaft,* 123, 30, 1983.
32. **Windisch, S., Kowalski, S., and Zander, I.,** Dough-raising tests with hybrid yeasts, *Eur. J. Appl. Microbiol.*, 3, 213, 1976.

Chapter 9

REPEATED BATCH FERMENTATION OF ETHANOL USING IMMOBILIZED CELLS OF *SACCHAROMYCES CEREVISIAE* IN A FLUIDIZED BIOREACTOR SYSTEM

A. Margaritis, D. te Bokkel, and M. El-Kashab

TABLE OF CONTENTS

ABSTRACT

Saccharomyces cerevisiae cells were immobilized by entrapment within small Ca-Alginate beads of about 2 mm in diameter and were used to ferment a glucose defined medium. A novel 26-ℓ fluidized bioreactor system was designed and built, and used with the immobilized yeast cells. Gentle fluidization of the Ca-Alginate beads and uniform mixing were achieved by means of internal jets and two concentric draft tubes which stabilized internal liquid patterns within the bioreactor. This bioreactor system facilitated easy removal of the CO_2 gas produced and was used in repeated batch mode of operation where the same immobilized yeast cells were used for a total of ten repeated batch cycles using fresh liquid medium at the beginning of each cycle. Due to high immobilized cell concentrations per unit bioreactor volume, very fast rates of ethanol fermentation were achieved. Using a 15% w/v glucose defined medium, the repeated batch fermentations using immobilized yeast cells were completed in only 4 to 5 hr giving ethanol to glucose yields about 98% of the theoretical maximum yield. These rates and yields of immobilized cells fermentations were much higher than those obtained from traditional fermentation systems employing free yeast cells. A maximum volumetric ethanol productivity of 21.2 g ETOH per liter per hour was obtained using immobilized yeast cells which is about 10 times higher than the typical volumetric ethanol productivity of 2 g ETOH per liter per hour obtained with free cell systems employed in traditional ethanol fermentation technology.

I. INTRODUCTION

Fermentation ethanol is rapidly emerging as an important renewable liquid fuel and/or octane booster for unleaded gasoline-ethanol mixtures. Present ethanol fermentation technology is obsolete and energy inefficient and involves the use of freely suspended yeast cells in batch operation mode requiring large-size fermentors. Typical batch fermentation times for a 10% sugar solution are in the order of 24 to 48 hr, carried out in ordinary mechanically stirred tanks.

In this paper, we report fermentation results using a novel fluidized bioreactor system with immobilized yeast cells which are trapped within small Ca-Alginate beads of about 2-mm diameter. High concentrations of immobilized cells per unit bioreactor volume are possible, which result in high volumetric ethanol productivity.

II. MATERIALS AND METHODS

A. The 26-ℓ Fluidized Bioreactor System

Figure 1 shows a schematic diagram of the 26-ℓ fluidized bioreactor used with immobilized cells of *Saccharomyces cerevisiae* trapped within Ca-Alginate beads of 2-mm diameter. In order to operate the bioreactor as a liquid fluidized system, a suction manifold was installed through the base of the bioreactor and connected to the suction side of a Duro® little giant pump. The jets with 2.0-mm orifices were connected to a manifold on the discharge side of the pump. Mixing was induced in the bioreactor by withdrawing fluid from the bioreactor through the suction manifold and pumping it back into the reactor through the jets. This resulted in the circulation patterns shown in Figure 1 where the heavy dots represent the Ca-Alginate beads which circulate up and down the bioreactor between the draft tubes. This arrangement facilitated the easy removal of CO_2 gas during fermentation during the gentle fluidization of Ca-Alginate beads containing immobilized yeast cells.

B. Yeast Cell Immobilization Procedure

S. cerevisiae Fleishman's came from the supplier in two forms, either as a yeast cake or

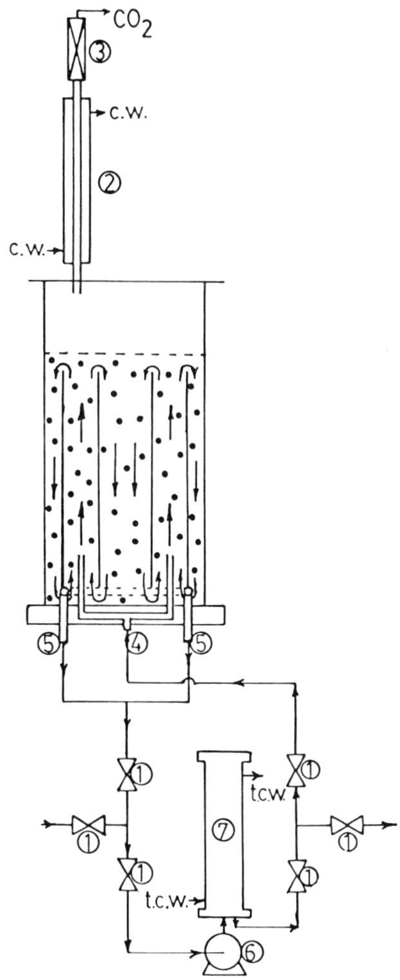

FIGURE 1. Schematic of 26-ℓ bioreactor configuration for liquid fluidized operation. (1) Control valves; (2) condensor; (3) exhaust gas filter; (4) inlet to jets; (5) outlet of suction manifold; (6) Duro® centrifugal pump; (7) temperature-control heat exchanger: c.w./cooling water, t.c.w./temperature-controlled water.

dried yeast granules. The dry weight per unit mass of each of these forms was determined by drying a sample at 105°C for 24 hr. The required concentration of yeast was revived according to the manufacturer's instructions. The cell mass was prepared in 1 to 2 ℓ of distilled water. An equal volume of 32 g/ℓ Na-Alginate was prepared in distilled water. When both prepared solutions were uniformly hydrated and well mixed, they were added together and gently mixed to uniformity. The resultant solution of 1.6% Na-Alginate and cells constituted the feed of the immobilization procedure. The hardening solution was 40 to 60 g/ℓ calcium chloride in water.

The 2-mm-diameter beads were generated using a multisyringe-needle manifold. It consisted of 30 26-gauge syringe needles mounted in a sealed housing. The feed solution (Na-Alginate plus cells) was pumped into the device and extruded through the syringe needles, and the droplets formed at the tips of the syringe needles fell into the $CaCl_2$ solution and were hardened. The beads were hardened overnight by keeping them in the $CaCl_2$ solution at 4°C and were then used in the fluidized bioreactor to ferment a glucose medium to ethanol.

C. Media and Fermentation Conditions

The medium used for the production of ethanol from glucose by immobilized cells of *S. cerevisiae* was initially 150 g/ℓ D-glucose, 12.9 g/ℓ yeast extract, 0.2 g/ℓ $MgSO_4 \cdot 7H_2O$, 2.0 g/ℓ NH_4Cl, and 0.1 g/ℓ $CaCl_2$ adjusted to pH = 5. After running the first batch experiment with 2-mm beads, it was noted that 60 to 70 g/ℓ ethanol was inhibitory to *S. cerevisiae*,[2] so glucose concentration was reduced to 100 g/ℓ. It had also previously been noted in our laboratory that 1 to 2 g/ℓ $CaCl_2$ stabilized immobilized cell beads and, therefore, the medium was supplemented with $CaCl_2$.

Experiments indicated that the addition of antifoam to the medium appeared to enhance CO_2 removal from the fermentor and prevented cavitation on the suction side of the liquid recycle pump. The resultant medium had the following composition for the repeated batch fermentation using 2-mm beads of immobilized cells: 100 g/ℓ D-glucose, 12.9 g/ℓ yeast extract, 0.2 g/ℓ $MgSO_4 \cdot 7H_2O$, 2.0 g/ℓ NH_4Cl, 2.0 g/ℓ $CaCl_2$, and 0.27 mℓ/ℓ antifoam B.

The omission of yeast extract from the medium for the last four repeated batches above showed that the immobilized yeast required less yeast extract. The yeast extract concentration was halved for all following experiments. The fermentation was carried out at 35°C using the heat exchanger and the temperature controller to maintain this temperature.

D. Analytical Methods

Glucose or reducing sugar concentrations were determined using the DNS (dinitrosalicylic acid) test of Miller[3] with minor modification of the DNS reagent concentrations: DNS, 10 g/ℓ; Na_2SO_3, 0.5 g/ℓ; NaOH, 10 g/ℓ; sodium potassium tartrate (Rochelle salt), 200 g/ℓ; and phenol, 2 g/ℓ. Ethanol analysis was done using a Varian® 3700 gas chromatograph.

Free biomass concentration was determined using the optical density (O.D.) method. A standard curve of biomass concentration (dry weight) and O.D. at 650 mm was prepared. The fermentor-free biomass samples were read against media blanks and the free biomass concentration determined from the standard curve.

Changes in immobilized cell concentration within the 2-mm beads were determined by dissolving six beads in 10 mℓ of 1% sodium tripolyphosphate and determining the cell concentration by the O.D. method. A hemocytometer cell-counting chamber was used to check changes in the total number of cells within the beads. Any changes in cell count were directly proportional to changes in cell concentration (grams dry weight) determined by the O.D. method.

Viable cell concentration in the beads was determined by dissolving beads in 1% sodium tripolyphosphate and selectively staining nonrespiring (usually dead) cells with Loeffler's methylene blue.[1,4] Live respiring cells actively exclude the stain. A hemocytometer cell-counting chamber slide was used to count the number of nonstained actively respiring cells and stained cells. From these values the percentage of viable (respiring) cells could be calculated. The total concentration of viable cells then was calculated from the initial bead loading.

III. RESULTS AND DISCUSSION

A. Batch Fermentation with Immobilized Cells

The results of a batch fermentation using 2-mm Ca-Alginate beads of immobilized cells of *S. cerevisiae* are shown in Figure 2. The bioreactor contained 13.58 g dry wt of immobilized yeast per liter of total working volume. The yeast viability was 74%. Therefore, the active biomass concentration was 10.05 g/ℓ. The total working volume of the bioreactor was 10 ℓ, of which 2 ℓ was the bead volume in the bioreactor of 8 ℓ of medium.

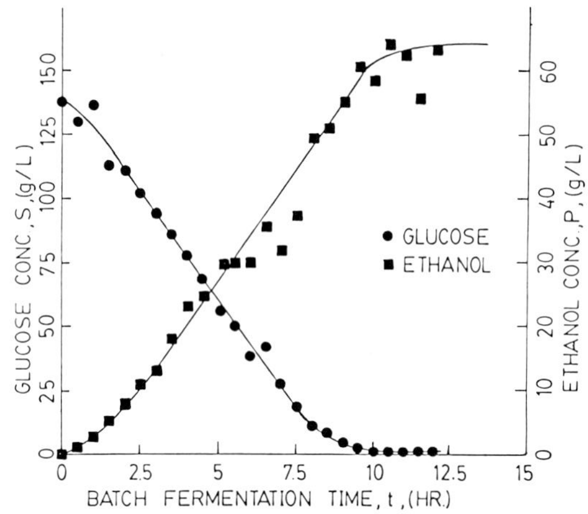

FIGURE 2. Results of a batch fermentation of glucose by 2-mm beads containing immobilized cells of *S. cerevisiae* in the 26-ℓ liquid fluidized bioreactor.

Complete fermentation of the 13.7% w/v initial glucose concentration occurred in about 8 hr. Typical batch fermentation using free (nonimmobilized) yeast cells took about 48 to 72 hr for complete fermentation. The lag phase normally present in free cell fermentation was absent in the immobilized yeast fermentation, as shown by the rapid initial glucose uptake rate in Figure 2. The high immobilized cell concentration in the bioreactor resulted in high rates of ethanol production and glucose consumption. The maximum rate of ethanol production was 7.08 g/ℓ/hr. Glucose was consumed at a maximum rate of 16.66 g glucose per liter per hour. The overall ethanol yield was 0.472 g ethanol per gram glucose utilized, which was 92.4% of the theoretical yield. The maximum ethanol concentration was 64 g/ℓ.

B. Repeated Batch Fermentation Using Immobilized Cells

After the completion of the batch run shown in Figure 2, the same Ca-Alginate beads containing immobilized yeast were used repeatedly to carry out ten separate batch fermentation runs. At the completion of each batch fermentation run, the liquid phase was removed while retaining the Ca-Alginate beads, and fresh glucose medium was charged into the fermentor to start the new run. The repeated batch fermentations were done using 2-mm beads containing 15.5 g dry weight immobilized yeast cells per liter of total bioreactor working volume with a viable cell concentration of 11.6 g dry weight cells per liter total bioreactor volume. The bead and medium volumes were 2 and 8 ℓ, respectively. The initial glucose and ethanol concentrations varied for each repeated batch fermentation. This was due to residual media left in the pump lines and associated with the beads after emptying the medium from the bioreactor when each batch was complete.

The results of these repeated batch fermentations are shown in Figures 3A, B, and C. These figures show the following parameters plotted as a function of the fermentation time: glucose concentration, ethanol concentration, free biomass in the liquid phase, and pH. The majority of the repeated batches completed glucose utilization in 4 to 5 hr.

Table 1 summarizes the ethanol yields and glucose utilization of each of the batches. The highest ethanol yields were attained in the last runs, and were 98% of the theoretical maximum yield of 0.51 g ethanol per gram glucose utilized. Glucose utilization in these fermentations was usually 99% of the original glucose. These excellent ethanol yields and percent sugars utilization were possible, because of the high immobilized yeast concentrations used per

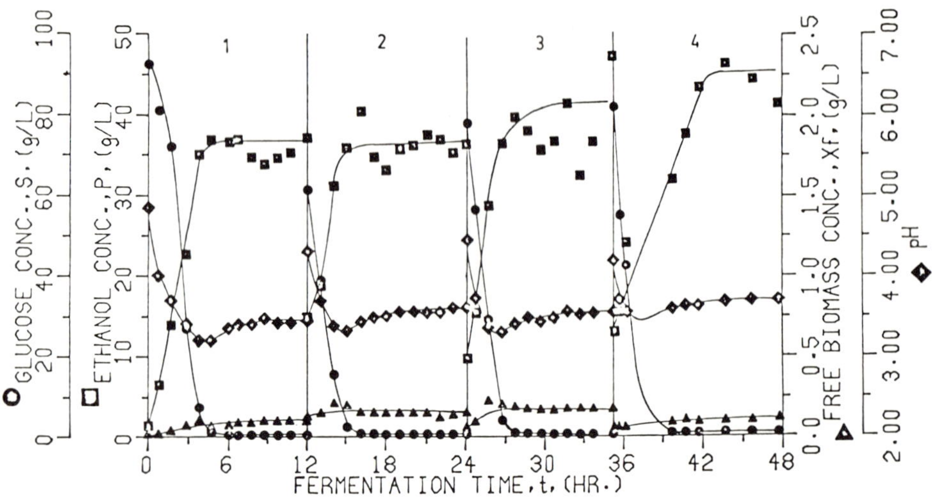

A

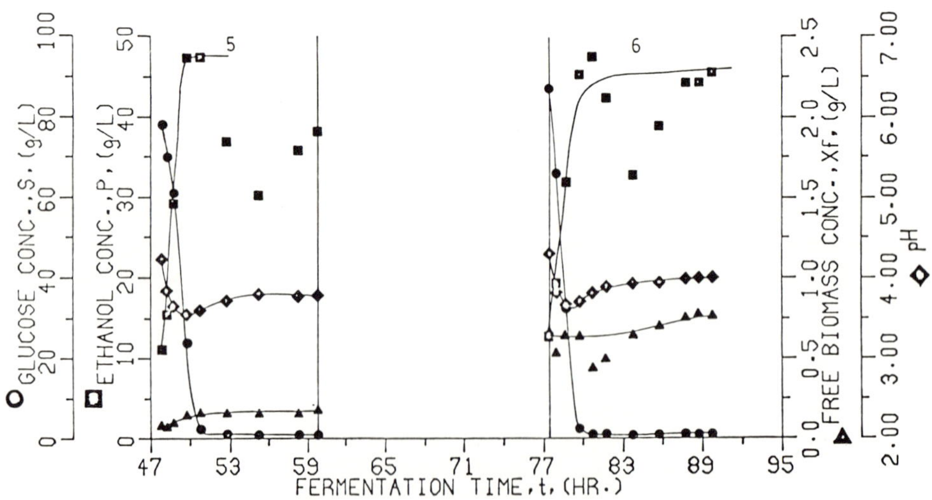

B

FIGURE 3. (A) Repeated batch fermentation for the first four batches using immobilized cells of *S. cerevisiae* in a fluidized bioreactor system. (B) Repeated batch number 5, 17-hr period with no fermentation, and batch number 6 using immobilized yeast cells in a fluidized bioreactor. (C) Repeated batch fermentations number 7 to 10 using immobilized yeast cells in a fluidized bioreactor.

unit bioreactor volume. The ethanol volumetric productivities increased for the first five batch runs and stabilized over runs 6 and 7. The maximum volumetric ethanol productivity was 21.2 g ETOH per liter per hour obtained during batch number 5. Typical batch volumetric ethanol productivities using free (nonimmobilized) yeast cells in traditional fermentation technology used today range between 2 and 4 g ETOH per liter per hour. This is primarily due to the low free cell concentrations used in these fermentations as opposed to our immobilized cell fermentation where the concentrations are very high. Therefore, the maximum volumetric ethanol productivities obtained with our immobilized cell system are about five to ten times higher than the typical productivities obtained with free cell systems employed in traditional ethanol batch fermentation technology.

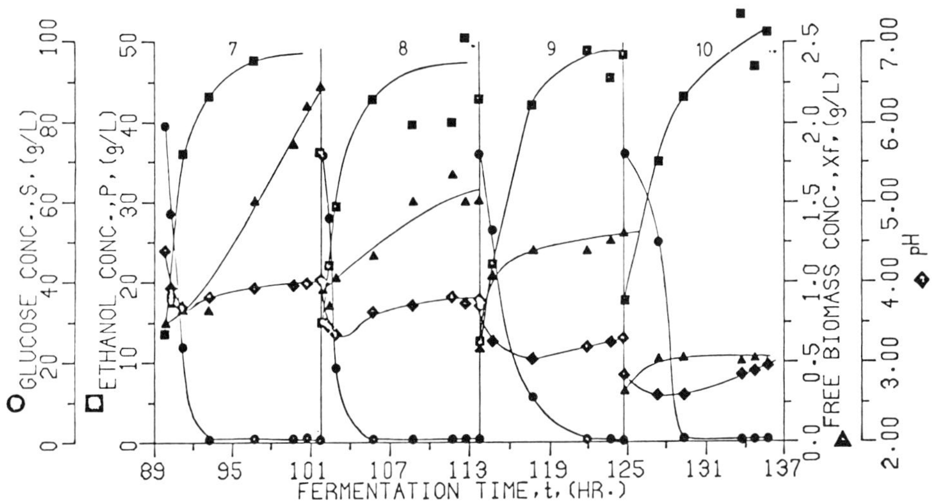

FIGURE 3C

Table 1
RESULTS OF REPEATED BATCH FERMENTATION USING RECYCLED IMMOBILIZED CELLS OF *S. CEREVISIAE*

Batch run no.	Ethanol yield $Y_{P/S}$ $\left(\frac{\text{g EtOH}}{\text{g glu}}\right)$	Ethanol yield % Theoretical ($Y_{P/S} \div 0.511$)	Glucose utilized (%)	Ethanol productivity $\frac{dp}{dt}$ (g/ℓ/hr)
1	0.409	80.0	99.5	9.42
2	0.422	82.6	99.1	8.66
3	0.387—0.485	75.7—94.5	99.05	10.02
4	0.412	80.7	98.95	12.20
5	0.480	93.97	98.71	21.2
17-Hr Gap between Batches				
6	0.414	81.0	98.67	14.1
7	0.442—0.510	86.5—100	98.66	15.66
8	0.499	97.7	98.83	14.43
9	0.505	98.9	99.17	10.2
10	0.500	97.9	99.19	6.67

During the 17-hr gap between runs 5 and 6, the immobilized cells were kept in the reactor with fresh media and no circulation. The yeast extract from the medium used in runs 8, 9, and 10 was not added to the medium. This resulted in reduced levels of free biomass production, lower ethanol productivities, and decreased fermentation rates with each succeeding batch.

The same Ca-Alginate beads containing immobilized cells of *S. cerevisiae* were used for 137 hr. During this time, the biomass loading in the beads increased to 2.8 times the original bead biomass loading.

ACKNOWLEDGMENTS

This work was supported by the National Research Council of Canada Contract Research DSS No. OSU82-00006.

REFERENCES

1. **Lee, S. S., Robinson, F. M., and Wang, H. Y.,** Rapid determination of yeast viability, *Biotechnol. Bioeng. Symp.*, 11, 641, 1981.
2. **Maiorella, B., Blanch, H. W., and Wilke, C. R.,** By-product inhibition effects on ethanolic fermentation by *Saccharomyces cerevisiae*, *Biotechnol. Bioeng.*, 25, 103, 1983.
3. **Miller, G. L.,** Use of dinitrosalicyclic acid reagent for determination of reducing sugar, *Anal. Chem.*, 31, 426, 1959.
4. **Trevors, J. T., Merrick, R. L., Russell, I., and Stewart, G. G.,** A comparison of methods for assessing yeast viability, *Biotechnol. Lett.*, 5, 131, 1983.

Chapter 10

FILAMENTOUS FUNGI: HOSTS FOR THE EXPRESSION OF FOREIGN PROTEINS

Alan Upshall and G. L. McKnight

TABLE OF CONTENTS

I. INTRODUCTION

The search for an efficient expression-secretion system capable of large-scale production of the high-dollar-value therapeutic metabolites is rapidly gaining momentum. Although *Saccharomyces cerevisiae* has achieved considerable popularity as a eukaryotic expression host and has been successfully employed to produce and secrete some foreign proteins,[1-3] it is well known that difficulties have been experienced with expression levels and with the ability of the organism to accurately process some types of proteins.[4] These problems have stimulated the investigation of other expression systems, of which potentially there are many. *S. cerevisiae* is one of many yeast species and, as has been well illustrated at this conference, a significant proportion of these are efficient at producing various metabolites. It is a fundamental feature and complexity of microbial taxonomy that different species within the groups of organisms show very diverse metabolic phenotypes. Among the yeasts, therefore, could be hidden a species which has all of the attributes desired for the production of heterologous proteins. The initial problem is identifying that species, closely followed by the requirement to develop molecular, genetic, and culture procedures. The successful entry of *S. cerevisiae* as an expression host into the commercial biotechnology market was primarily a consequence of its excellent genetics and molecular biology. Other yeast species have neither of these attributes. A similar appraisal can be made for other groups of organisms that have been used, or are being considered for use, as production strains, among which mammalian cell cultures, plant cell cultures, Bacaulovirus, and filamentous fungi probably constitute the main contenders. It is not our intention to embark on a comparison of the relative merits of these systems; time will tell whether any one will prove to be the ubiquitous expression host. One fact will prevail: the commercial producers will employ the most economical process for the production of the approved product. The inclusion of filamentous fungi is not simply because the molecular technology is now available, nor because successful expression of active prochymosin has been achieved in *Aspergillus*,[5] but because the new molecular developments supplement the methods and information already available for the industrial species. This article concentrates on this group.

Historically, filamentous fungi, like some yeasts, have long been employed for the production of a wide variety of commercially desirable metabolites.[6,7] They therefore present the attractive attribute of having in place a strong industrial base with considerable fermentation technology and supportive biochemical and physiological experience. The various species have been manipulated by mutagenic treatments which, when combined with advances in fermentation technology, have enabled the exploitation of mutants which yield extremely high levels of product. This method of strain development has been practiced for the production of secondary metabolites such as antibiotics (e.g., Calam et al.[8]) and primary metabolites such as industrial enzymes,[9] and the molecular technology should impact these two areas.[10] For the purposes of the expression of foreign proteins, however, the strains used for producing industrial enzymes command immediate attention, since they are potentially ready-made hosts, with the proviso that they can be satisfactorily manipulated.

The strain development laboratories of the producing companies have large culture collections representing every important stage of the pedigree of the strains they have employed or are currently employing. These collections are a wealth of genetic heterogeneity and provide a firm base for the exploitation of filamentous fungi as expression hosts. However, four basic problems can be identified:

1. These collections are generally available only to appropriate company employees, which virtually eliminates their contribution to the development of suitable high level expression systems by the emerging companies, even under contract agreements.
2. Most strains are generally characterized only in respect of the trait for which they were

selected and for vigor in specific growth conditions. Other observations on phenotype are incidental.

3. Industrial species have been minimally studied from a genetic point of view and have poorly developed methods of analysis.
4. The exploitation of these species as hosts requires the development of appropriate molecular methods.

Despite these difficulties, what makes this segment of the industry so important is that some of the mutations involved in the steps to increase titer and activity of crude enzyme preparations are potentially in genes which are important components of an expression system, namely, secretion and protease production. For example, elimination or reduction of the levels of extracellular protein degrading enzymes would allow a culture to yield greater quantities of active protein; mutations could enhance the production of processing enzymes that are necessary to release active enzyme. It follows that concomitant increased production of the proform of a secreted enzyme, accompanied by increased levels of a signal peptide cleavage protein, might allow increased yield of active product. Essentially, strains with a combination of all of the positive factors required for high level production, as would probably be isolated following the successive mutagenesis regimen, would be ideal expression hosts. This firm industrial base is now being supplemented by a new understanding of cellular processes gained from the application of genetic, molecular, and biochemical methods. The remainder of this article will summarize and comment on those areas which are important to the production and secretion of heterologous proteins.

II. MOLECULAR METHODS

With the lack of genetic understanding of the industrial species, the development of the molecular technology must rely on the technical and genetic systems evolved for the model species *Aspergillus nidulans* and *Neurospora crassa.* The validity of extrapolating from one species to another is supported by observations from several investigations. Genetic similarities are shown by the patterns of mutant recovery for resistance to potassium chlorate and the phenotypes and groupings of the mutants. These are essentially the same in *Penicillium chrysogenum,*[11] *A. niger* and *A. terreus,*[52] and *N. crassa,*[12] indicating a common regulation of nitrate assimilation and metabolism. Transformation experiments have shown the ability of genes from one species to function in another. For example, the *amd*S[13] and *arg*B[14] genes of *A. nidulans* will transform strains of *A. niger* and of *A. terreus;*[53] the *pyr*-4 gene of *N. crassa* efficiently transforms *A. nidulans.*[15,16] the *amd*S function has allowed for the development of a transformation system for wild-type strains of *A. niger* and *A. terreus*, although the efficiencies of transformation are low. Undoubtedly, these efficiencies will improve, but even at the low level they provide access to molecular manipulation of industrial genomes. In all systems, stable transformants generally are the outcome of the integration of the DNA into the chromosome,[17,18] which, in the case of the transformation of *A. niger* by the *amd*S gene, occurs in the absence of any detectable homology between the DNA molecules in Southern blot analysis.[13] Evidence for freely replicating plasmids in transformants has been obtained in *N. crassa.*[19] In the two model species, molecular manipulation has become refined with the development of gene disruption and gene replacement techniques,[20,21] and there is no reason to believe that these methods will not work in other fungal species when their molecular and genetic base is better developed.

To exploit filamentous fungi as expression hosts in a controlled manner requires isolating components of expression cassettes, namely, promoter sequences from regulated genes and from constitutively expressed genes, terminator sequences, and protein secretion signals if the product is desired to be extracellular. Many regulated systems have been characterized, for example,[22-24] and advantage of this understanding has been taken to obtain cDNA clones unique to induced cultures.[25,26] The corresponding genes are sources of promoter and terminator sequences and, in the case of glucoamylase genes,[27,28] cellobiohydrolase genes[29] and glucosidase gene[30] also provide fungal secretion signal peptides. These methods do not allow for the isolation of cDNA copies of mRNA from constitutively expressed genes, which are present in both induced and uninduced cultures. The approach to cDNA cloning taken by McKnight and colleagues,[31] which is a modification of functional cDNA selection first described by Okayama and Berg,[32] overcomes this deficiency. In this method, cDNA molecules are cloned between a *S. cerevisiae* promoter and terminator sequence present in a yeast vector. Transformation of appropriate yeast mutants automatically selects for full-length cDNA molecules. In addition to this feature, this method (1) exploits the wealth of genetic understanding and selectable phenotypes of *S. cerevisiae*; (2) avoids promoter function and intron splicing problems, since it has been shown that foreign promoter sequences function inefficiently, at best, in *S. cerevisiae*[31,33] and that introns not carrying yeast-specific consensus sequences cannot be spliced out of mRNA transcripts;[31,33] (3) employs relatively simple and efficient yeast transformation and selection procedures and benefits from all of the advantages of the high copy number plasmid nature of yeast transformants. Since the selection is by functional complementation, this method can be used to isolate cDNA clones from transcripts which are not necessarily the most prevalent within the extracted mRNA population.

III. SECRETION

The basic commercial requirement for any heterologous protein is the economic production of pure active product, and the ability to extract the product from the culture medium is a significant consideration in the cost equation. Extraction and purification is easiest if the culture medium is simple, containing only the essential nutrients, rather than one which is made up of complex organic components. Commercially employed species are usually able to grow on defined minimal media, but in their natural environment they are heterotrophic, and they secrete a wide variety of degradative enzymes in order to obtain their essential nutrients, e.g., proteases,[34,35] carbohydrases,[36] lipases,[37] and nucleases.[38,39] The production of most of these enzymes is regulated either by derepression followed by substrate induction, as with carbohydratase enzymes,[40] or by derepression in the absence or depletion of a nutrient source.[41] While the kinetics of the production and properties of many of these enzymes have been determined, in only a few have the genetic basis of the regulation and/or molecular analysis of the gene been undertaken. Probably the most thoroughly studied example is the glucoamylase gene of *A. niger* and its taxonomically very close relative, *A. awamori*.[27,28,42] By comparing the translated nucleotide sequence of this gene with the amino acid sequence of the mature protein, the leader peptide of the preform of the enzyme responsible for directing the protein through the secretory pathway was identified. This signal peptide has features similar to other prokaryotic and eukaryotic signal sequences[43,44] which allowed the accurate prediction of the site of cleavage of the signal peptide from the primary protein product. Further, a second endopeptidase processing site after the arginine residue of a dibasic lys-arg couplet was detected.[42] The ability of the glucoamylase signal peptide to be processed and to direct secretion from the yeast[33] and the apparent commonality of prokaryotic and eukaryotic processing signals[45] suggest that the glucomylase signal peptide and probably also the signal sequences of other secreted proteins may be suitable secretion signals for heterologous protein expression in filamentous fungi. However, given these functional similarities, the alternative approach to secretion may be to use the natural secretion signal of a secreted mammalian protein for expression in the fungal host. This could simplify the construction of the appropriate expression units and allow the integrity of the amino acid composition of the protein to be easily retained.

IV. GLYCOSYLATION

Some of the commercially important secreted mammalian proteins are glycosylated by the addition of significant levels of D-mannose sugars covalently bound to asparagine-linked *N*-acetyl-*d*-glucosamine molecules.[47] The fungal enzymes that have been characterized are also glycosylated and most show the same pattern of glycosylation as higher eukaryotes.[47] Some proteins show additional carbohydrate molecules O-linked to serine or threonine.[28] The importance of the glycosylation is apparently different depending on the enzyme studied. The antibiotic tunicamycin disrupts glycosylation, and in studies with secreted glucosidases of *A. fumigatus*, reduced extracellular levels and activity of these proteins were observed in cultures grown in the presence of the drug.[47] Further, Wasserman and Hultin[48] have shown that the removal of the carbohydrate moieties from *A. niger* catalase increases the sensitivity of the enzyme to protease activity. In contrast, the secretion of other glycoproteins is not impaired by exposure to tunicamycin, suggesting that in these cases, the glycosylation pattern was not a factor in directing the secretion of the molecule,[49] although it was possible that the structure of the added carbohydrate was insensitive to tunicamycin. In either case, it is evident that filamentous fungi can modify and process secreted proteins in a fashion similar, if not identical, to higher eukaryotes,[47,50] and should therefore provide good hosts in these respects. Additionally, these observations give filamentous fungi an extra flexibility in that it will be practical to construct expression units such that the coding sequence of the heterologous protein can be fused in translational reading frame to the signal peptide of several different secreted proteins. This versatility will allow the best combination of signal peptide (be it fungal or natural) and glycosylation pattern to be achieved for maximal expression.

V. PROTEASES

Filamentous fungi secrete large quantities of proteases, a trait to be expected from heterotrophic organisms which absorb mineral nutrients following the extracellular digestion of nutrient sources. Commercially, protease production uses *A. oryzae*. Proteases pose serious problems for the production of heterologous proteins and it is imperative that their presence and devastating effects be controlled, if not eliminated. Under the controlled artificial culture conditions of nutrient sufficiency, the expression of the extracellular proteases is not required for the generation of essential nutrients, and hence, the elimination of their activity would not be detrimental to growth.

Two basic approaches are available: first, genetic modification by inducing mutations which reduce extracellular protease levels; second, to culture the organism under conditions in which protease production is repressed.[41] Much is known about the occurrence, properties, and environmental regulation of production. Various secreted proteases have been identified[35] and production of all forms is generally in response to the depletion of nitrogen or sulfur from the culture medium.[34,35] With the current sophisticated methods of control of fermentor growth conditions, precise control of the constituents of the medium is a realistic practicality, and under such circumstances, extracellular proteases should present only a minor problem. Intracellular proteases may be more difficult to control. These are important for correct processing of proteins and are therefore not dispensable. They can achieve significance because of autolysis of hyphae in culture. Rigid monitoring of production conditions would minimize the potential dangers. Attempts to define mutations in genes involved in protease production and secretion have proven unsuccessful.[35] One mutation identified in *A. nidulans*, designated *xpr*D1, confers constitutive production of extracellular protease and constitutive expression of other gene loci which are normally repressed by ammonium.[41] Subsequent studies have shown this mutation to be an allele of the *are*A locus, a transacting regulatory gene involved in ammonium repression.[51] The enhanced production of *xpr*D1 mutant strains, therefore, is a consequence of the elimination of ammonium regulation rather than a mutation in one of the genes coding for a specific protease. Clearly, starting from a wild-type strain, the formal genetic or molecular genetic approach to defining proteases would be a long-term project, and this is an area where a high yielding industrial strain or one of its progenitors, which might be protease deficient, could present considerable advantages for production purposes. In their absence, control of protease would initially have to involve empirical manipulation of the medium to ensure growth of the strains under conditions which repress protease production.

VI. CONCLUSION

The information available suggests that filamentous fungi could prove ideal hosts for expressing therapeutic metabolites on a commercial scale. However, the full and rapid exploitation of this group of organisms will depend on the speed with which the molecular methods can be applied to industrial species and the development of a basic understanding of their genetic architecture. Although not addressed in this brief overview, it is clear that, in addition to this feature, the application of the molecular methodology could offer new avenues for the improvement of those strains already employed for producing valuable fungal metabolites, by incorporating molecular cloning and gene manipulation into a strain improvement program. Further, the ability to clone genes from one species for expression in another greatly expands the genetic resources available, which should lead to an even greater exploitation of filamentous fungi than has been achieved to date.

REFERENCES

1. **Singh, A., Lugovoy, J. M., Kohr, W. J., and Perry, L. J.,** Synthesis, secretion and processing of α-factor-interferon fusion proteins in yeast, *Nucl. Acids Res.*, 12, 8927, 1984.
2. **Brake, A. J., Merryweather, J. P., Coit, D. G., Heberlein, U. A., Masiarz, F. R., Mullenbach, G. T., Urdea, M. S., Valenzuela, P., and Barr, P. J.,** α-Factor directed synthesis and secretion of mature foreign proteins in *Saccharomyces cerivisiae, Proc. Natl. Acad. Sci. U.S.A.*, 81, 4642, 1984.
3. **Bitter, G. A., Chen, K. K., Banks, A. R., and Lai, P. H.,** Secretion of foreign proteins from *Saccharomyces cerivisiae* directed by α-factor gene fusions, *Proc. Natl. Acad. Sci. U.S.A.*, 81, 5330, 1984.
4. **Smith, R. A., Duncan, M. J., and Moir, D. T.,** Heterologous protein secretion from yeast, *Science*, 229, 1219, 1985.
5. *Genetic Engineering News*, 9, 1, 1985.
6. **Turner, W. B.,** *Fungal Metabolites*, Academic Press, London, 1971.
7. **Ball, C.,** Filamentous fungi, in *Genetics and Breeding of Industrial Microorganisms*, Ball, C., Ed., CRC Press, Boca Raton, Fla., 1983, 159.
8. **Calam, C. T., Daglish, B. L., and McCann, E. P.,** Penicillin: tactics in strain improvement, in *Genetics of Industrial Micro-Organisms*, Macdonald, K. D., Ed., Academic Press, London, 1976, 273.
9. **Blain, J. A.,** Industrial enzyme production, in *The Filamentous Fungi*, Vol. 1, Smith, J. E. and Berry, D. R., Eds., Arnold, London, 1974, 193.
10. **Upshall, A.,** Filamentous fungi in biotechnology, *Biotechniques*, 4, 158, 1986.
11. **Birkett, J. A. and Rowlands, R. T.,** Chlorate resistance and nitrate assimilation in industrial strains of *Penicillium chrysogenum, J. Gen. Microbiol.*, 123, 281, 1981.
12. **Dunn-Coleman, N. S., Tomsett, A. B., and Garrett, R. H.,** The regulation of nitrate assimilation in *Neurospora crassa:* biochemical analysis of the nmr-1 mutants, *Mol. Gen. Genet.*, 182, 234, 1981.
13. **Kelly, J. M. and Hynes, M. J.,** Transformation of *Aspergillus niger* by the *amd*S gene of *Aspergillus nidulans, EMBO J.*, 4, 475, 1985.
14. **Buxton, F. P., Gwynne, D. I., and Davies, R. W.,** Transformation of *Aspergillus niger* using the *arg*B gene of *Aspergillus nidulans, Gene*, 37, 207, 1985.
15. **Ballance, D. J., Buxton, F. P., and Turner, G.,** Transformation of *Aspergillus nidulans* by the orotidine-5-phosphate decarboxylase gene of *Neurospora crassa, Biochem. Biophys. Res. Commun.*, 112, 284, 1983.

16. **Ballance, D. J. and Turner, G.,** Development of a high frequency transforming vector for *Aspergillus nidulans, Gene,* 36, 321, 1985.
17. **Yelton, M. E., Hamer, J. E., and Timberlake, W. E.,** Transformation of *Aspergillus nidulans* by using a *trp*C plasmid, *Proc. Natl. Acad. Sci. U.S.A.,* 81, 1470, 1984.
18. **Upshall, A.,** Genetic and molecular analysis of *arg*B$^+$ transformants of *Asperillus nidulans, Curr. Genet.,* 10, 593, 1986.
19. **Grant, D. M., Lambowitz, A. M., Rambosek, J. A., and Kinsey, J. A.,** Transformation of *Neurospora crassa* with recombinant plasmids containing the cloned glutamate dehydrogenase (am) gene: evidence for autonomous replication of the transforming plasmid, *Mol. Cell. Biol.,* 4, 2041, 1984.
20. **Miller, B. L., Miller, K. Y., and Timberlake, W. E.,** Direct and indirect gene replacements in *Aspergillus nidulans, Mol. Cell. Biol.,* 5, 1714, 1985.
21. **Paietta, J. V. and Marzluf, G. A.,** Gene disruption by transformation in *Neurospora crassa, Mol. Cell. Biol.,* 5, 1554, 1985.
22. **Arst, H. N. and Bailey, C. R.,** The regulation of carbon metabolism in *Aspergillus nidulans,* in *Genetics and Physiology of Aspergillus,* Smith, J. E. and Pateman, J. A., Eds., Academic Press, London, 1977, 131.
23. **Arst, H. N.,** Regulation of gene expression in *Aspergillus nidulans, Microbiol. Sci.,* 1, 137, 1984.
24. **Marzluf, G. A.,** Regulation of nitrogen metabolism and gene expression in fungi, *Microbiol. Rev.,* 45, 437, 1981.
25. **Hynes, M. J., Corrick, C. M., and King, J. A.,** Isolation of genomic clones containing the *amd*S gene of *Aspergillus nidulans* and their use in the analysis of structural and regulatory mutations, *Mol. Cell. Biol.,* 3, 1430, 1983.
26. **Lockington, R. A., Sealy-Lewis, E. M., Scazzoccio, C., and Davies, R. W.,** Cloning and characterization of the ethanol utilization regulon in *Aspergillus nidulans, Gene,* 33, 137, 1985.
27. **Boel, E., Hjort, I., Svensson, B., Norris, F., Norris, K. E., and Fiil, N.,** Glucoamylases G1 and G2 from *Aspergillus niger* are synthesized from two different but closely related mRNAs, *EMBO J.,* 5, 1097, 1984.
28. **Nunberg, J. H., Meade, J. H., Cole, G., Lawyer, F. C., McCabe, P., Schweickhardt, V., Tal, R., Wittman, V. P., Flatgaard, J. E., and Innes, M. A.,** Molecular cloning and characterization of the glucoamylase gene of *Aspergillus awamori, Mol. Cell. Biol.,* 2, 2306, 1984.
29. **Shoemaker, S., Schweikhardt, V., Ladner, M., Gelfand, D., Kwok, S., Myambo, K., and Innes, M.,** Molecular cloning of exo-cellobiohydrolase I derived from *Trichoderma reesei* strain L27, *Biotechnology,* 1, 691, 1984.
30. **Pentillae, M. E., Nevalainen, K. M. H., Raynal, A., and Knowles, J. K. C.,** Cloning of *Aspergillus niger* genes in yeast. Expression of the gene coding *Aspergillus* β glucosidase, *Mol. Gen. Genet.,* 194, 494, 1984.
31. **McKnight, G. L., Kato, H., Upshall, A., Parker, M. D., Saari, G., and O'Hara, P. J.,** Identification and molecular analysis of a third *Aspergillus nidulans* alcohol dehydrogenase gene, *EMBO J.,* 4, 2093, 1985.
32. **Okayama, H. and Berg, P.,** High efficiency cloning of full length cDNA, *Mol. Cell. Biol.,* 2, 161, 1982.
33. **Innes, M. A., Holland, M. J., McCabe, P. C., Cole, G. E., Wittman, V. P., Tal, R., Watt, W. K., Gelfand, D. H., Holland, J. P., and Meade, J. H.,** Expression, glycosylation and secretion of an *Aspergillus* glucoamylase by *Saccharomyces cerevisiae, Science,* 228, 21, 1985.
34. **Tomonaga, G.,** Preferential synthesis of extracellular protease by *Aspergillus niger* in sulfur deficiency, *J. Gen. Appl. Microbiol.,* 12, 267, 1966.
35. **Cohen, B. L.,** The proteases of *Aspergilli,* in *Genetics and Physiology of Aspergillus,* Smith, J. E. and Pateman, J. A., Eds., Academic Press, London, 1977, 281.
36. **Speake, B. K., Malley, D. J., and Hemming, F. W.,** The effect of tunicamycin on secreted glycosidases of *Aspergillus niger, Arch. Biochem. Biophys.,* 210, 110, 1981.
37. **Aunstrup, K., Andersen, O., Falch, E. A., and Neilsen, T. K.,** Production of microbial enzymes, in *Microbial Technology,* Vol. 1, Peppler, H. J. and Perlman, D., Eds., Academic Press, New York, 1979, 282.
38. **Campbell, A. M. and Winder, F. G.,** Properties of deoxyribonuclease 4 from *Aspergillus nidulans, Biochem. Biophys. Acta,* 746, 125, 1983.
39. **Kafer, E. and Witchell, G. R.,** Effects of *Neurospora* nuclease halo (*nuh*) mutants on secretion of two phosphate repressible alkaline deoxyribonucleases, *Biochem. Genet.,* 22, 403, 1984.
40. **Smith, J. E. and Berry, D. R.,** *The Filamentous Fungi,* Vol. 2, Edward Arnold, London, 1975.
41. **Cohen, B. L.,** Regulation of intracellular and extracellular neutral and alkaline proteases in *Aspergillus nidulans, J. Gen. Microbiol.,* 79, 311, 1973.
42. **Boel, E., Hansen, M. T., Hjort, I., Hoegh, I., and Fiil, N. P.,** Two different types of intervening sequences in the glucoamylase gene from *Aspergillus niger, EMBO J.,* 3, 1581, 1984.

43. **von Heijne, G.,** Patterns of amino acids near signal-sequence cleavage sites, *Eur. J. Biochem.,* 133, 17, 1983.
44. **von Heijne, G.,** How signal species maintain cleavage specificity, *J. Mol. Biol.,* 173, 243, 1984.
45. **Perlman, D. and Halvorson, H. O.,** A putative signal peptidase recognition site and sequence in eukaryotic and prokaryotic signal peptides, *J. Mol. Biol.,* 167, 391, 1983.
46. **Hemming, F. W.,** Dolichol phosphate, a co-enzyme in the glycosylation of animal membrane bound glycoproteins, *Biochem. Soc. Trans.,* 5, 1223, 1977.
47. **Elbein, A. D., Mitchell, M., and Molyneux, R. J.,** Effect of castanospermine on the structure and secretion of glycoprotein enzymes in *Aspergillus fumigatus, J. Bacteriol.,* 160, 67, 1985.
48. **Wasserman, B. P. and Hultin, H. O.,** Effect of deglycosylation on the stability of *Aspergillus niger* catalase, *Arch. Biochem. Biophys.,* 212, 385, 1981.
49. **Struck, D. K., Suita, P. B., Lane, M. D., and Lennarz, W. J.,** Effect of tunicamycin on the secretion of serum proteins by primary cultures of rat and chick hepatocytes, *J. Biol. Chem.,* 253, 5332, 1978.
50. **Weintraub, B. D., Stannard, B. S., Linnekin, D., and Marshall, M.,** Relationship of glycosylation to *de novo* thyroid stimulating hormone biosynthesis and secretion by mouse pituitary tumor cells, *J. Biol. Chem.,* 255, 5715, 1980.
51. **Arst, H. N.,** *Fungal Systems in Eukaryotic Genes,* Maclean, N., Gregory, S. P., and Glavell, R. A., Eds., Butterworths, Oxford, 1983, 433.
52. **Upshall, A.,** unpublished.
53. **Upshall, A. and Favreau, M. A.,** unpublished.

Chapter 11

USE OF THE MELIBIASE PROMOTER AND SIGNAL PEPTIDE TO EXPRESS A BACTERIAL CELLULASE FROM YEAST

Nigel A. Skipper, Richard P. Bozzato, Daniel Vetter, R. Wayne Davies, Raymond Wong, and James E. Hopper

TABLE OF CONTENTS

I. INTRODUCTION

Much effort is being directed to using *Saccharomyces cerevisiae* as a host for the engineered expression of proteins. Particular attention is being paid to genetically engineering the secretion of these proteins. In addition to the obvious benefit of simple recovery of the product, the purpose of engineering secretion is to take advantage of posttranslational events such as glycosylation and endoproteolysis and to allow for the formation of correct secondary structures during the secretory process. Examples of proteins shown to be secreted from *S. cerevisiae* by the expression of genetically engineered DNA sequences include human interferon,[1-3] β-endorphin,[3] yeast invertase,[4] wheat α-amylase,[5] human epidermal growth factor,[6] fungal glucoamylase,[7] calf prochymosin and bovine growth hormone,[8] mouse interleukin 2,[9] and a bacterial cellulase.[10] In most cases expression was from autonomously replicating plasmids in which a suitable yeast promoter was fused, via a sequence encoding a yeast signal peptide, to the gene of interest. In one example, the operative gene fusion was also examined in an integrated state and found to be more effective in this state than in a plasmid.[8]

We are developing a secretion system which takes advantage of the properties of the *S. carlsbergensis* MEL1 gene. This gene encodes melibiase (α-galactosidase, E.C. 3.2.1.22), a secreted glycoprotein. Expression of MEL1 in *S. cerevisiae* is regulated by the carbon source via the GAL4 and GAL80 gene products, and under inducing conditions, melibiase can constitute over 40% of the extracellular protein in cultures.[11,12] We have undertaken a systematic examination of the molecular basis of the regulation and expression of extracellular melibiase. One approach is to determine the role of the melibiase signal peptide in secretion by assessing its effects on expression of novel coding sequences. In this paper we report on results of some preliminary experiments which have used the melibiase promoter and signal peptide to effect expression of a bacterial cellulase from yeast.

II. CONSTRUCTION OF EXPRESSION VEHICLES

A. Rationale

DNA sequencing of the MEL1 gene and amino acid sequencing of the secreted melibiase protein have demonstrated that the secreted protein is the product of a cleavage of an 18-amino acid amino-terminal peptide from the primary gene product, premelibiase.[12] The 18-residue preregion is, presumably, a signal peptide which directs the nascent protein to the yeast secretory pathway. The 5′ flank of the premelibiase sequence contains a putative upstream activation site, or UAS_M, based on homology to the UAS_G known to exist in a similar position upstream of the similarly regulated GAL1 and GAL10 genes.[13-15] The UAS_M is thought to be the site at which carbon source regulation of the MEL1 promoter is exerted, via the products of the GAL4 and GAL80 regulatory genes. Our strategy in constructing expression vehicles based on MEL1 was to first generate a portable DNA fragment or cassette containing both the promoter (including regulatory sites such as UAS_M) and the sequence for the melibiase signal peptide, the latter being provided with an adjacent cloning site suitable for in-frame fusion to our test gene. Following insertion of the gene, the cassette could then be placed into a suitable yeast plasmid for regulated expression of the test protein as an extracellular product of yeast growth.

B. Cassette Construction

Plasmid pMP550 is Yep24 containing a functional MEL1 gene within a 4.3-kb BamHI-HindIII fragment.[11] The 4.3-kb fragment contains a unique SphI site within the sequence encoding the melibiase signal peptide.[12] By a series of cloning steps, including two which involved synthetic oligonucleotides, a derivative of pMP550, plasmid p4, was constructed

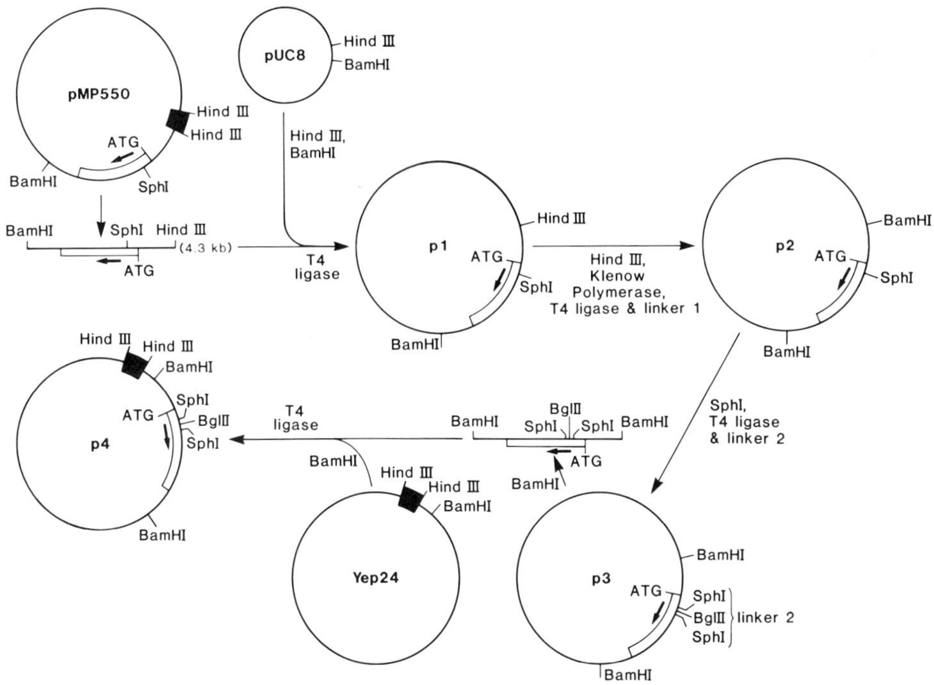

FIGURE 1. Construction of expression cassette. Standard methods[24] were used to manipulate plasmids and restriction fragments. The method of Lathe et al.[25] was used to insert linkers 1 and 2 into plasmids. The linkers were synthesized using phosphotriester chemistry and had the following sequence:

```
(linker 1), 5' CCGGATCCGG
            3' GGCCTAGGCC
(linker 2), 5'     CATCAGTTTGAAGGGCGTTTTTGGGAAGATCTCTGCATG
            3' GTACGTAGTCAAACTTCCCGCAAAAACCCTTCTAGAGAC
```

Key: ▬, URA3; ▭, premelibiase structural gene; →, direction of transcription.

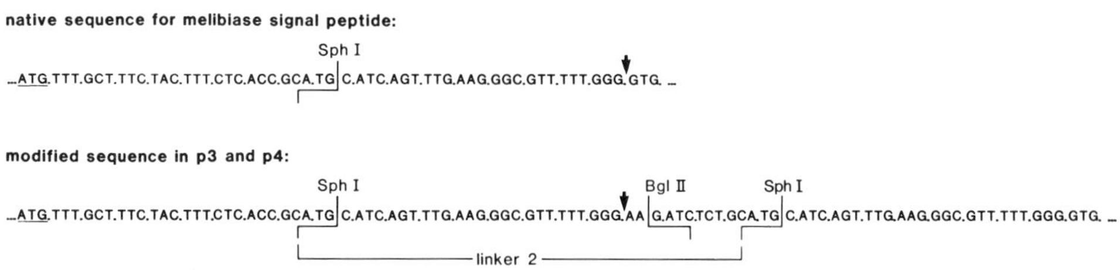

FIGURE 2. Nucleotide sequence of the native melibiase signal peptide[12] and of the modified sequence in plasmids p3 and p4 (Figure 1). The modified sequence was confirmed by dideoxy sequencing[26] using a synthetic oligonucleotide as a primer. The arrow marks the glycine codon (GGG) which immediately precedes the amino terminus of mature (secreted) melibiase.[12]

(Figure 1). This plasmid contains the 4.3-kb fragment bounded by BamHI sites and contains a unique BglII cloning site adjacent to the melibiase signal peptide codons (Figures 1 and 2). The 4.3-kb fragment constitutes the MELI expression cassette.

C. Melibiase: Cellulase Fusion

The soil bacterium *Cellulomonas fimi* ATCC 484 produces several extracellular cellu-

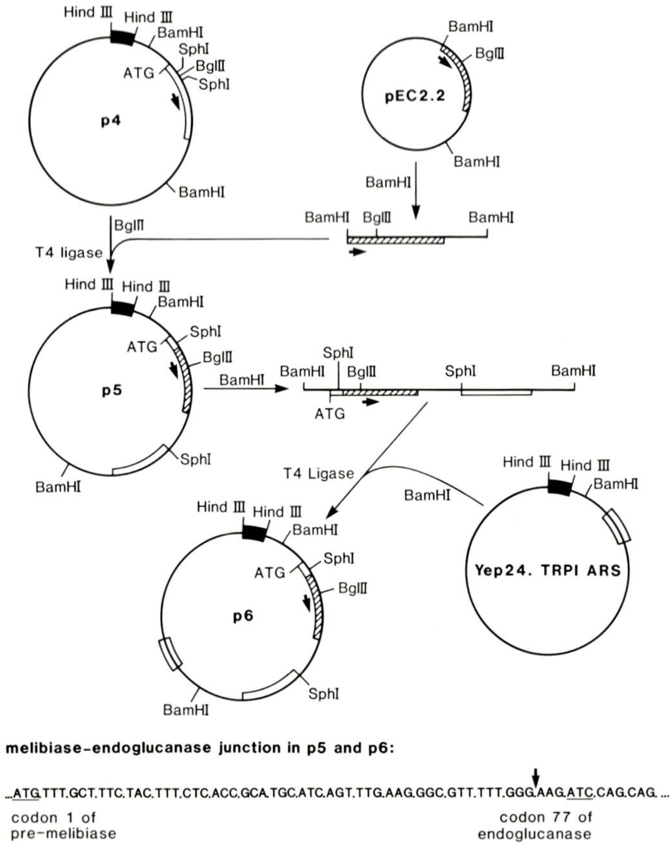

FIGURE 3. Construction of a melibiase/endoglucanase gene fusion in plasmid Yep24.TRP1 ARS, and nucleotide sequence at the junction of the fusion. The nucleotide sequence was determined as described in Figure 2. The arrow marks the glycine codon (GGG) which immediately precedes the amino terminus of mature (secreted) melibiase.[12] Plasmid Yep24.TRPI ARS was constructed by cloning the 1.4-kb EcoRI fragment from Yrp7, as a filled-in fragment, into the filled-in SalI site of Yep24. ▬, URA3; ▨, endoglucanase coding sequence; ▭, premelibiase structural gene; ⊟, TRP1 ARS.

lases,[16] including endoglucanases (1,4-[1,3:1,4]-β-D-glucan-4-glucanohydrolase, E.C.3.2.1.4). Plasmid pEC2.2[10] is pBR322 containing part of the coding sequence for one of these endoglucanases, within a 2.4-kb BamHI fragment (Figure 3). The endoglucanase sequence lacks the codons for the first 76 amino terminal residues of the enzyme, including the entire bacterial signal peptide.

The 2.4-kb BamHI fragment has been shown previously to encode an active, extracellular endoglucanase in yeast when expressed from a plasmid (pK 2.4) in which it was fused to the sequence for the amino terminus of the yeast K1 preprotoxin, downstream of the alcohol dehydrogenase I promoter.[10] It was used here to test the utility of the MEL1-based expression system.

A melibiase/endoglucanase gene fusion was generated by cloning the 2.4-kb BamHI fragment from pEC2.2 into the BglII site of plasmid p4. One of the plasmids with the insert in the correct orientation (p5, Figure 3) was then cut with BamHI to retrieve the gene fusion on a 6.7-kb fragment, and this was cloned into the BamHI site of Yep 24.TRPI ARS to generate plasmid p6 (Figure 3). The DNA sequence at the melibiase/endoglucanase junction in p5 and p6 is shown in Figure 3; it encodes the native melibiase signal peptide, a lysine residue, then the endoglucanase gene starting at residue 77.

III. EXPRESSION OF ENDOGLUCANASE BY YEAST

S. cerevisiae strain 20B12 (*trp1 pep4.3*)[17] was used to examine expression of endoglucanase from plasmid p6; it does not produce endogenous endoglucanase activity.[10] Competent cells were prepared by the LiCl procedure,[18] then transformed to tryptophan independence with plasmid p6. Cells were then assayed for extracellular endoglucanase by the Congo red staining procedure[19] following growth on trp-selective medium containing 2% agar, 2% glucose, and 0.9% carboxymethylcellulose.[10] Carboxymethylcellulose hydrolysis by endoglucanase is seen as a clear zone against a background which stains red. About 10% of the colonies tested produced extracellular activity by this test. The remainder evidently were false transformants, since plasmid could not be detected in them, either by attempted plasmid rescue into *Escherichia coli* RRI or by assay for the plasmid-encoded penicillinase by a published method.[20] The cause of the apparent high frequency of *trp1* to TRP1 reversion in our stock of strain 20B12 is not known.

One endoglucanase-positive isolate, transformant 130.1, was grown in batch culture and assayed over a growth cycle for endoglucanase using the quantitative assay of Miller et al.,[21] which uses a dinitrosalicyclic acid (DNS) reagent to measure release of reducing sugar equivalents from carboxymethylcellulose. For comparison, endoglucanase produced by strain 20B12 containing plasmid pK2.4[10] (transformant 43.10) was assayed in a parallel culture. In both cases extracellular activity was produced in parallel to cell number, although the cells containing the MEL1-based plasmid produced consistently more activity than those containing pK2.4 (Figure 4). Microscopic examination provided no evidence of cell damage in these cultures, therefore, the endoglucanase appeared to be a true product of secretion.

IV. EFFECT OF A PUTATIVE UP-SECRETION MUTATION

A. Isolation of Up-Secretion Mutants

It was known from previous work[10] that strain 20B12 produces a very low level of extracellular endoglucanase when transformed with plasmid pL5.19, a plasmid in which the *C. fimi* endoglucanase coding sequence is present as an essentially direct fusion to the yeast alcohol dehydrogenase I promoter, without an intentional yeast signal sequence. 20B12 containing pL5.19 gives only very faint clearing on carboxymethylcellulose plates, allowing variants that produce higher levels of extracellular endoglucanase activity to be detected as colonies giving stronger cleared zones. A transformant containing pL5.19 was irradiated with UV light, incubated in the dark for 60 min, then plated onto agar-containing trp-selective medium, 2% glucose, and 0.9% carboxymethylcellulose. Colonies were stained with Congo red and those that produced a particularly bright colorless zone were selected. These isolates were cured of pL5.19, by growth in nonselective medium, then transformed with plasmid pK2.4 to tryptophan independence and assayed by the DNS reagent for endoglucanase activity in order to show that the genetic alteration lay in the host cell rather than in the endoglucanase gene or pL5.19. Many putative up-secretion mutants have been obtained in this way and are currently being studied. The mutant used in the work reported here, 20B12.DV3, produced from three- to fourfold more activity per cell from plasmid pK2.4 than does the parent strain, 20B12.

B. Effect on Secretion from the Melibiase-Based Vectors

Strain 20B12.DV3 was transformed to tryptophan independence with plasmid p6 (Figure 3) and assayed for endoglucanase production. A typical isolate (transformant 131.3) produced

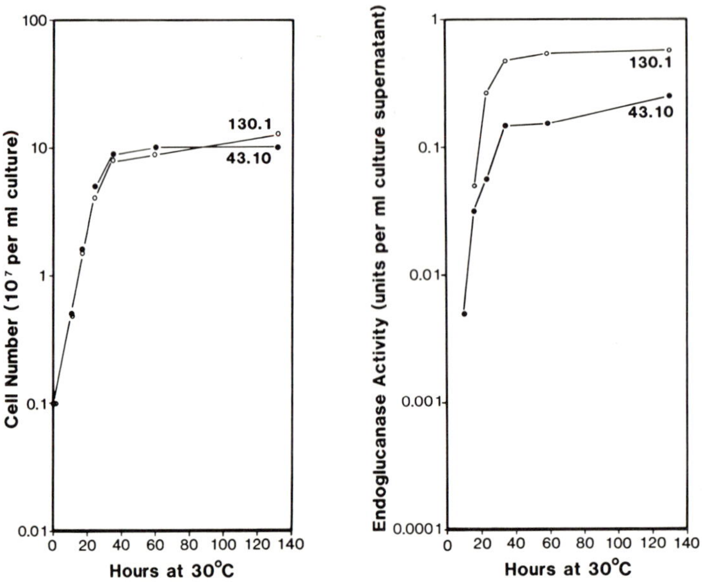

FIGURE 4. Endoglucanase expression by growing yeast cultures. Yeast transformants 130.1 and 43.10 (see text) were grown at 30°C in trp-selective medium[10] containing 2% lactic acid, 3% glycerol, 0.05% glucose, and 2% galactose as the carbon source; pH was 6.8 to 7.0. At intervals the cultures were sampled for cell number and for extracellular endoglucanase activity. Endoglucanase was assayed with the DNS reagent using carboxymethylcellulose as substrate, as described,[10] using culture supernatant samples that had been dialyzed, lyophilized, and reconstituted in water. One unit produces an absorbance at 530 nm of 1.0 in 1 hr at 37°C.

from two- to threefold more activity per cell than did strain 20B12 containing the same plasmid (transformant 130.1) (Table 1), a result essentially identical to that seen with plasmid pK2.4 in which the promoter and signal peptide used to express the enzyme are quite different. Since others[8] have reported that up-secretion mutations can affect the yield of more than one protein, we asked whether melibiase yield is affected in the same way as endoglucanase in strain 20B12.DV3. Plasmid p7 (Figure 5) was constructed for this purpose; it contains the MEL1 gene from plasmid p2 (Figure 1) inserted into the shuttle vector Yep24.TRPI ARS. Strains 20B12 and 20B12.DV3 produced essentially identical amounts of extracellular melibiase per cell from plasmid p7 (Table 1; transformants 127.1 and 128.2). This apparently specific effect of the mutation in 20B12.DV3 on endoglucanase yield has several possible explanations. One possibility is that melibiase secretion is saturating the secretory pathway in strain 20B12 so that no yield increase is possible.

V. REGULATED EXPRESSION

A. Introduction

Melibiase is produced at a low basal level, unless specific inducing or repressing carbon sources are present in culture media. The presence of galactose induces melibiase activity 20- to 25-fold,[22] while glucose represses it to an almost undetectable level.[23] Northern analysis has shown that this regulation is reflected by the level of melibiase mRNA, i.e., it appears to be largely a result of regulation of the promoter activity.[11] This regulation is mediated by the products of the GAL4 and GAL80 genes: GAL4 is required for both basal melibiase activity and for galactose induction of activity, while GAL80 antagonizes the action of

Table 1
EXPRESSION OF ENDOGLUCANASE AND MELIBIASE IN TRANSFORMANTS OF STRAINS 20B12 AND 20B12.DV3[a]

				Extracellular endoglucanase activity		Extracellular melibiase activity	
Strain	Plasmid	Transformant	Carbon source	m U/mℓ	m U/10^8 cells	m U/mℓ	m U/10^8 cells
20B12	p6	130.1	NI	5.21	2.61		
			I	15.49	8.15		
			R	16.66	5.95		
	p7	128.2	NI			5,400	1,680
			I			10,600	3,310
			R			13,200	2,640
20B12.DV3	p6	131.3	NI	14.58	16.2		
			I	39.12	27.9		
			R	34.96	9.2		
	p7	127.1	NI			3,700	1,850
			I			8,400	4,200
			R			4,900	1,360

[a] Cells were grown to stationary phase at 30°C in trp-selective medium[10] at either pH 6.8 to 7.0 (transformants 130.1, 131.3) or pH 4.6 to 4.8 (transformants 128.2, 127.1). The medium contained carbon sources either noninducing (NI), inducing (I), or repressing (R) for the melibiase promoter. Cultures were then assayed for either extracellular endoglucanase (Figure 3, legend) or extracellular melibiase,[12] and for cell number. One endoglucanase unit produces 1 μmol reducing sugar per minute at 37°C. One unit of melibiase produces 1 μmol *p*-nitrophenol per minute at 30°C using *p*-nitrophenylgalactopyranoside as substrate. The noninducing carbon source was 2% lactic acid, 3% glycerol, 0.05% glucose; the inducing carbon source was noninducing components plus 2% galactose; the repressing carbon source comprised the inducing components plus 2% glucose.

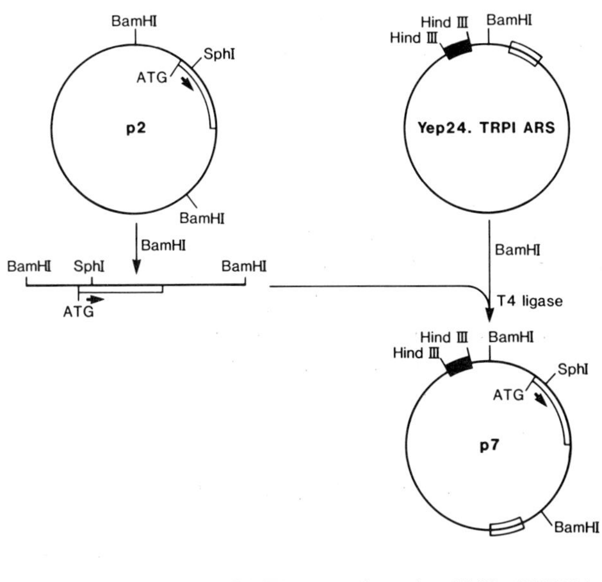

FIGURE 5. Insertion of MEL1 gene into plasmid Yep24.TRP1 ARS. ▬ , URA3; ▭ , premelibiase structural gene; ⊟ , TRP1 ARS.

GAL4.[11] Galactose overcomes this inhibitory action of GAL80. The basis of glucose repression is not understood.

B. Carbon Source Regulation in Yeast Transformants

Cultures of strains 20B12 or 20B12.DV3 containing plasmid p6 were assayed for extracellular endoglucanase activity following growth in media containing carbon sources either noninducing, inducing, or repressing for the melibiase promoter. In parallel, cultures containing plasmid p7 were grown and assayed for extracellular melibiase activity. The results are shown in Table 1. In both strain 20B12 and strain 20B12.DV3 the expression of each enzyme responded to the carbon source in the way predicted, although the degree of regulation did not correspond to our expectations. We are currently measuring promoter activity more directly, by Northern analysis of mRNA levels, to further examine these results.

VI. CHARACTERIZATION OF YEAST ENDOGLUCANASE

A. Purification

Ion-exchange chromatography of the proteins contained in the supernatant fluids from a 5-ℓ yeast culture containing the endoglucanase resolved them into a complex pattern. The endoglucanase activity was contained within a major class of the proteins that eluted in 0.25 *M* NaCl (Figure 6). The pooled active fractions were then subjected to size-exclusion chromatography. Initial experiments showed that the activity behaved as a very high molecular weight species. It was suspected that the endoglucanase was binding to secreted yeast glycoproteins, due to its affinity for carbohydrates. Equilibration of the active protein from the ion-exchange column to the substrate analog 1-*O*-methyl-α-D-glycopyranoside allowed for its subsequent fractionation on size-exclusion columns into three active fractions (A, B, and C, Figure 7). Each of these fractions was electrophoresed under reducing conditions in polyacrylamide gels and stained with Coomassie Blue. Apparent molecular weights of 44, 97, and 150 kdaltons were assigned to fractions A, B, and C, based on this analysis (results not shown). Further analysis is needed to resolve the relationship of these species; it is possible that the two larger species are differently glycosylated variants of the 44-kdalton species. The 44-kdalton species correspond in size to the predicted endoglucanase encoded by the gene fragment used in our expression plasmids, provided that the enzyme is not glycosylated.

B. Yield

Melibiase is secreted at a high level for a *S. cerevisiae* protein; the activity produced in strain 20B12 under inducing conditions (transformant 128.2, Table 1) is equivalent to approximately 26 mg melibiase per liter, or 8.1 $\mu g/10^8$ cells. In contrast, endoglucanase production was low. We estimate that the activity produced in strain 20B12 under inducing conditions (transformant 130.1, Table 1) corresponds to 54 μg endoglucanase per liter, or 11 $ng/10^8$ cells. This yield was increased to about 136 μg/ℓ (36 $ng/10^8$ cells) in the mutant 20B12.DV3 (transformant 131.3, Table 1).

VII. CONCLUDING REMARKS

We have shown that a plasmid containing the melibiase promoter region and the sequence encoding the melibiase signal peptide can be used to express a bacterial endoglucanase as an extracellular yeast protein. While the present yield of the cellulase is low, we believe the melibiase-based expression sequences have potential for development into a commercially viable delivery system both for cellulases and for other useful proteins. This development will require a far better understanding of the determinants of protein secretion, particularly of the role of the mature (secreted) protein in its own passage through the secretory apparatus.

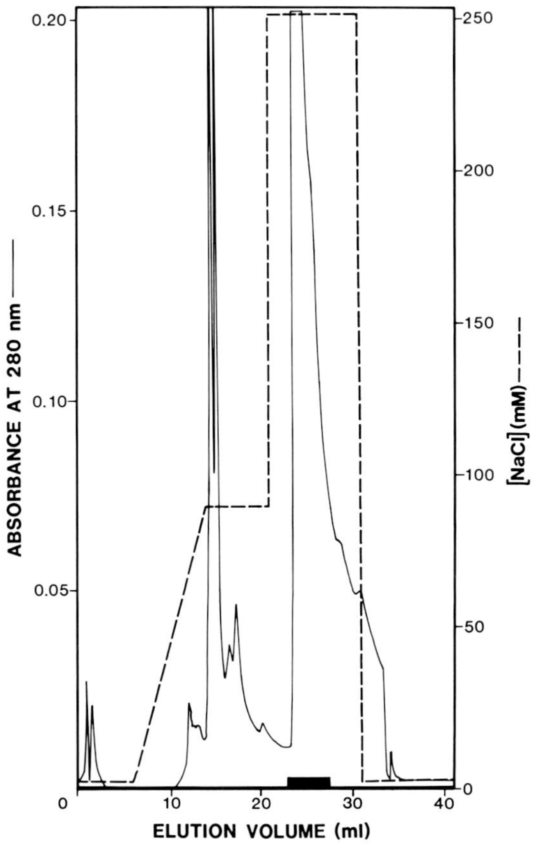

FIGURE 6. Ion-exchange chromatography of yeast endoglucanase. A total of 5 ℓ of culture was centrifuged to remove the cells and the supernatant fluids concentrated by ultrafiltration using an Amicon® PM-10 membrane. The concentrate was dialyzed against water and then against 20 m*M* piperazine, pH 11.0 (column buffer), to give a 10-mℓ sample containing 20 mg protein. The sample was applied to a Pharmacia® Mono Q HR 5/5 column, in 0.5-mℓ aliquots. Each aliquot was put through a chromatography cycle in which protein that bound to the column was eluted by application of 250 m*M* NaCl in column buffer; flow rate was 1.0 mℓ min^{-1}. Fractions were assayed for endoglucanase by dispensing a 15-μℓ sample onto 0.5-cm^2 Whatman® GF/C glass fiber filters on the surface of a petri dish containing 2% agar, 0.9% carboxymethylcellulose, and 50 m*M* sodium phosphate buffer, pH 7.0. Following a 15-hr incubation at 37°C the filters were removed and the dishes stained with Congo red. Fractions containing endoglucanase activity, seen as clear zones against a red background on the petri dishes, were pooled (indicated by a solid bar on the diagram).

ACKNOWLEDGMENTS

We thank Margaret Sutherland and Claudia Curry for their expert assistance, Richard Barnett and Heather Erfle for the oligonucleotides, Martin Sumner-Smith for his discussion, and Maureen Langsford, Tony Warren, Douglas Kilburn, and Robert Miller for providing us with the sequence of the amino terminus of the secreted *C. fimi* endoglucanase. This work was supported in part by the National Research Council of Canada under the Program for Industry/Laboratory Projects.

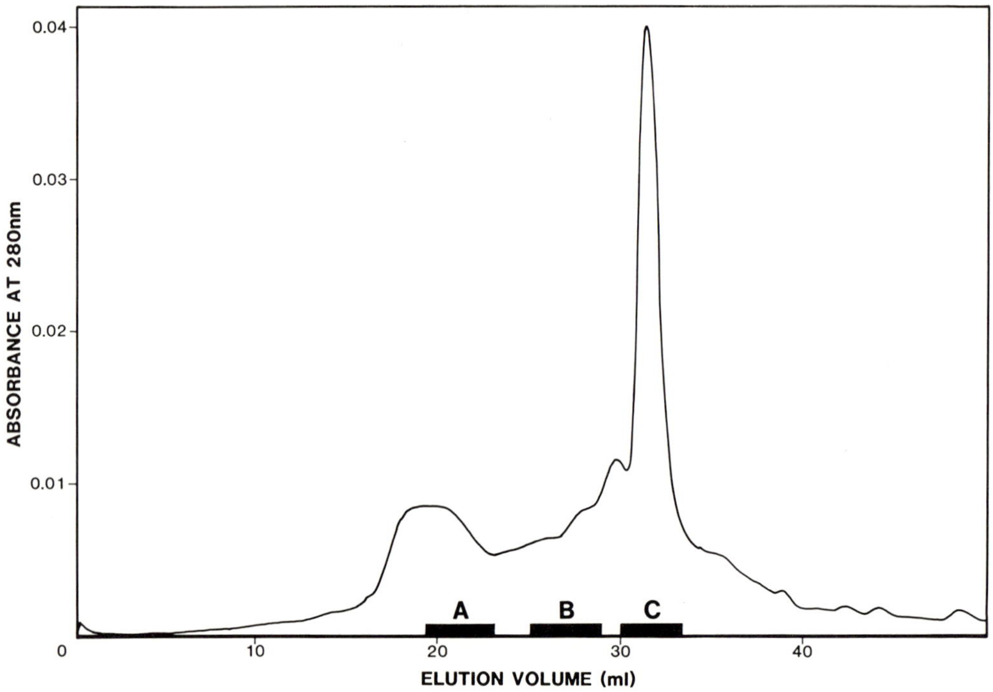

FIGURE 7. Size-exclusion chromatography of yeast endoglucanase. The pooled active fractions from the Mono Q column (Figure 6) were combined and concentrated to 0.5 mℓ. The sample was then dialyzed against 0.1 *M* Tris HC1 buffer, pH 7.5, containing 0.5 *M* 1-*O*-methyl-α-D-glucopyranoside, (column buffer), and 100-μℓ aliquots applied to a Pharmacia® Superose 12 column that was connected in series to a Pharmacia® Superose 6 column. Half milliliter fractions were collected at a flow rate of 0.25 mℓ.min^{-1} and assayed for endoglucanase as described in Figure 6. Three separate regions in the elution profile (solid bars on the diagram) contained activity; fractions comprising each region were pooled for further analysis (see text).

REFERENCES

1. **Hitzeman, R. A., Leung, D. W., Perry, L. J., Kohr, W. J., Levine, H. L., and Goeddel, D. V.,** Secretion of human interferons by yeast, *Science*, 219, 620, 1983.
2. **Singh, A., Lugovoy, J. M., Kohr, W. J., and Perry, L. J.,** Synthesis, secretion and processing of α-factor-interferon fusion proteins in yeast, *Nucleic Acids Res.*, 12, 6927, 1984.
3. **Bitter, G. A., Chen, K. K., Banks, A. R., and Lai, P.-H.,** Secretion of foreign proteins from *Saccharomyces cerevisiae* directed by α-factor gene fusions, *Proc. Natl. Acad. Sci. U.S.A.*, 81, 5330, 1984.
4. **Emr, S. D., Schekman, R., Flessel, M. C., and Thorner, J.,** An MFα1-SUC2 (α factor-invertase) gene fusion for study of protein localisation and gene expression in yeast, *Proc. Natl. Acad. Sci. U.S.A.*, 88, 7080, 1983.
5. **Rothstein, S. J., Lazarus, C. M., Smith, W. E., Baulcombe, D. C., and Gatenby, A. A.,** Secretion of a wheat α-amylase expressed in yeast, *Nature*, 308, 662, 1984.
6. **Brake, A. J., Merryweather, J. P., Coit, D. G., Heberlein, U. A., Masiarz, F. R., Mullenbach, G. T., Urdea, M. S., Valenzuela, P., and Barr, P. J.,** α Factor-directed synthesis and secretion of mature foreign proteins in *Saccharomyces cerevisiae*, *Proc. Natl. Acad. Sci. U.S.A.*, 81, 4642, 1984.
7. **Innis, M. A., Holland, M. J., McCabe, P. C., Cole, G. E., Wittman, V. P., Tal, R., Watt, K. W. K., Gelfand, D. H., Holland, J. P., and Meade, J. H.,** Expression, glycosylation, and secretion of an *Aspergillus* glucoamylase by *Saccharomyces cerevisiae*, *Science*, 228, 21, 1985.
8. **Smith, R. A., Duncan, M. J., and Moir, D. T.,** Heterologous protein secretion from yeast, *Science*, 229, 1219, 1985.

9. **Miyajima, A., Bond, M. W., Otsu, K., Arai, K., and Arai, N.,** Secretion of mature mouse interleukin-2 by *Saccharomyces cerevisiae:* use of a general secretion vector containing promoter and leader sequences of the mating pheromone α-factor, *Gene,* 37, 155, 1985.
10. **Skipper, N., Sutherland, M., Davies, R. W., Kilburn, D., Miller, R. C., Warren, A., and Wong, R.,** Secretion of a bacterial cellulase by yeast, *Science,* 230, 958, 1985.
11. **Post-Beittenmiller, M. A., Hamilton, R. W., and Hopper, J. E.,** Regulation of basal and induced levels of the MEL1 transcript in *Saccharomyces cerevisiae, Mol. Cell. Biol.,* 4, 1238, 1984.
12. **Sumner-Smith, M., Bozzato, R. P., Skipper, N., Davies, R. W., and Hopper, J. E.,** Analysis of the inducible MEL1 gene of *Saccharomyces carlsbergenesis* and its secreted product, alpha-galactosidase (melibiase), *Gene,* 36, 333, 1985.
13. **Johnston, M. and Davis, R.,** Sequences that regulate the divergent GAL1-GAL10 promoter in *Saccharomyces cerevisiae, Mol. Cell. Biol.,* 4, 1440, 1984.
14. **West, R., Jr., Yocum, R., and Ptashne, M.,** Yeast GAL1-GAL10 divergent promoter region: location and function of the upstream activating sequence-UAS_G, *Mol. Cell. Biol.,* 4, 2467, 1984.
15. **Yocum, R., Hanley, S., West, R., Jr., and Ptashne, M.,** Use of *lacZ* fusions to delimit regulatory elements of the inducible divergent GAL1-GAL10 promoter in *Saccharomyces cerevisiae, Mol. Cell. Biol.,* 4, 1985, 1984.
16. **Langsford, M. L., Gilkes, N. R., Wakarchuk, W. W., Kilburn, D. G., Miller, R. C., Jr., and Warren, R. A. J.,** The cellulase system of *Cellulomonas fimi, J. Gen. Microbiol.,* 130, 1367, 1984.
17. **Jones, E.,** Proteinase mutants of *Saccharomyces cerevisiae, Genetics,* 85, 23, 1977.
18. **Ito, H., Murata, K., and Kimura, A.,** Transformation of intact yeast cells treated with alkali cations or thiol compounds, *Agric. Biol. Chem.,* 48, 341, 1984.
19. **Teather, R. M. and Wood, P. J.,** Use of congo-red polysaccharide interactions in enumeration and characterisation of cellulolytic bacteria from the bovine rumen, *Appl. Environ. Microbiol.,* 43, 777, 1982.
20. **Chevallier, M. R. and Aigle, M.,** Qualitative detection of penicillinase produced by yeast strains carrying chimeric yeast-coli plasmids, *FEBS Lett.,* 108, 179, 1979.
21. **Miller, G. L., Blum, R., Glennon, W. E., and Burton, A. L.,** Measurement of carboxymethylcellulase activity, *Anal. Biochem.,* 2, 127, 1960.
22. **Buckholz, R. G. and Adams, B. G.,** Induction and genetics of two α-galactosidase activities in *Saccharomyces cerevisiae, Mol. Gen. Genet.,* 182, 77, 1981.
23. **Friis, J. and Ottohenghi, P.,** Localisation of melibiase in a strain of yeast, *C. R. Lab. Carlsberg,* 31, 272, 1959.
24. **Maniatis, T., Fritsch, E. F., and Sambrook, J.,** *Molecular Cloning. A Laboratory Manual,* Cold Spring Harbor Laboratory, Cold Spring Harbor, N.Y., 1982.
25. **Lathe, R., Kieny, M. P., Skory, S., and Lecocq, J. P.,** Linker tailing: unphosphorylated linker oligonucleotides for joining DNA termini, *DNA,* 3, 173, 1984.
26. **Sanger, F., Nicklen, S., and Coulson, A. R.,** DNA sequencing with chain-terminating inhibitors, *Proc. Natl. Acad. Sci. U.S.A.,* 74, 5463, 1977.

Chapter 12

EXPRESSION OF GLUCOAMYLASE IN YEAST FOR FERMENTATION OF LIQUIFIED STARCH

M. A. Innis, P. C. McCabe, G. E. Cole, V. P. Wittman, R. Tal, D. H. Gelfand, M. J. Holland, A. Ben-Bassat, J. McRae, D. Inlow, and J. H. Meade

TABLE OF CONTENTS

I. INTRODUCTION

Industrial processes using yeast for the production of ethanol from liquified starch can be improved by introducing a glucoamylase gene into *Saccharomyces cerevisiae*. Since the extracellular glucoamylase from *Aspergillus awamori* hydrolyzes starch at the α-1-4 and α-1-6 linkages to yield glucose, it is the enzyme of choice for commercial saccharification processes. We previously described the cloning and characterization of the *A. awamori* glucoamylase gene.[1] The glucoamylase gene was expressed in yeast and glucoamylase was secreted at levels sufficient to allow wild-type strains to grow on starch.[2] By expressing the glucoamylase gene from the yeast *ENO1* promoter contained on an autonomously replicating yeast vector, glucoamylase was secreted at levels of up to 1.2% of total cell protein, making it the major extracellular protein in the medium. A significant loss of up to 12% per generation of glucoamylase activity occurs when strains containing these plasmids were grown under nonselective as well as selective conditions. To eliminate this loss the glucoamylase gene was integrated into the genome of a wild-type industrial distiller's yeast at one to two copies per genome, and the effects on growth and fermentation were examined. During the growth on maltrin M-150 (a partially hydrolyzed starch preparation) about 93 to 95% of the starch was fermented. These fermentations had a higher ethanol and cell yield and a lower glycerol yield than of fermentation of comparable media with glucose. The expression and secretion of glucoamylase, per se, had no obvious deleterious effect on cell physiology. Also, there is no loss of glucoamylase activity during 800 generations under nonselective growth condition.

II. RESULTS

A. Expression Vectors and Yeast Strains

A. awamori and the related species *A. niger* secrete two glycosylated forms of glucoamylase, designated GAI and GAII. Both forms are encoded by a single gene, are immunologically related, and have the same amino-terminal sequence. The two forms differ in primary structure and length of their carboxyl-terminal sequence. These differences arise as a consequence of differential RNA splicing. Comparison of the primary structure of the glucoamylase gene with cDNA clones has established that the first four intervening sequences are spliced from both GAI and GAII mRNAs. A fifth intervening sequence is spliced from GAII.[3]

The expression vectors pAC1 or pMAC101 contain an *Escherichia coli* origin of replication, the *bla*, gene of pBR322, the yeast 2μ origin of replication, and a yeast *LEU* 2 structural gene. The vector also contains the promoter and terminator regions of the yeast enolase gene *ENO1*. A *Hind*III fragment containing the glucoamylase gene coding for GAI was inserted into pAC1. This plasmid was designated pGAC9.

Plasmids pPM16 and pPM18 are improved glucoamylase expression vectors, and plasmid pPM20 is an integrative glucoamylase expression vector. We have characterized glucoamylase expression in a laboratory yeast strain C468 (α *leu*2-2 *his* 3-11 his 3-15 *mal*) and an industrial yeast strain C162.

B. Expression and Secretion of Glucoamylase

To examine if transformed clones expressed and secreted glucoamylase, they were transferred onto plates containing 2% starch and checked for clearing of turbidity. After 6 days of growth at 30°C, C468:pGAC9 transformants, but not C468:pAC1 transformants, showed distinct zones of clearing around the growth area, indicating that they can hydrolyze starch (Figure 1). In addition, C468:pGAC9 transformants had a marked growth advantage on starch compared to transformants containing pAC1.

Transformants containing pMAC101 or pPM16 were grown in liquid cultures containing

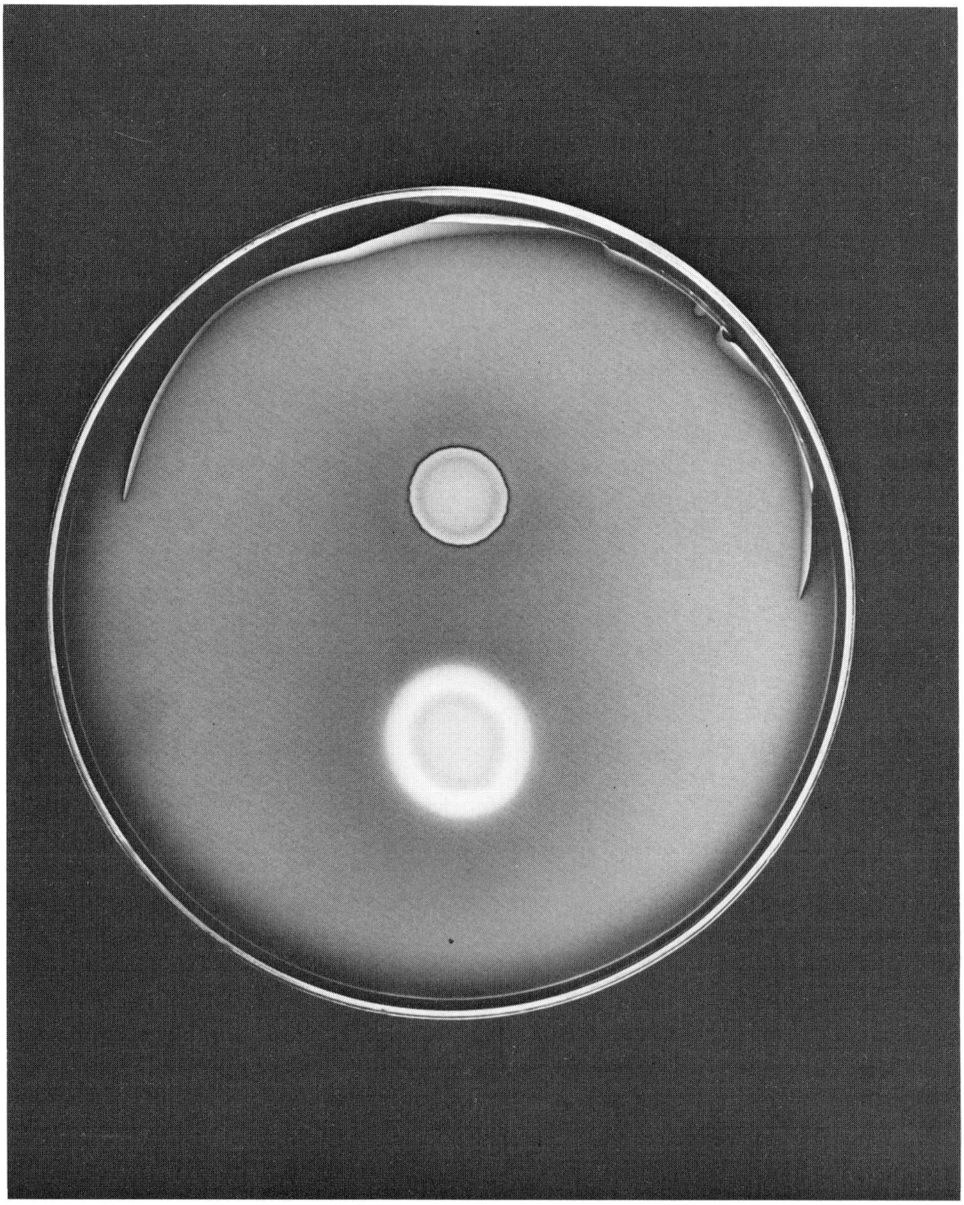

FIGURE 1. Starch clearing plate assay for glucoamylase expression. (Reprinted from Innis, M. A. et al., *Science*, 228(4695), 1985. With permission.)

starch and the secreted glucoamylase was analyzed on an SDS-urea polyacrylamide gel (7% acrylamide and 7.6 urea) (Figure 2). Glucoamylase activity was determined for extracellular supernatants of stationary phase cultures. Glucoamylase activity was determined by measuring glucose released from a buffered (pH 4.5) soluble starch solution at 37°C. We estimated that the extracellular glucoamylase constitutes 80 to 90% of the total extracellular proteins in the medium. The sample from C468:pMAC101 was fivefold more concentrated than C468:pPM18 in order to detect extracellular proteins other than glucoamylase.

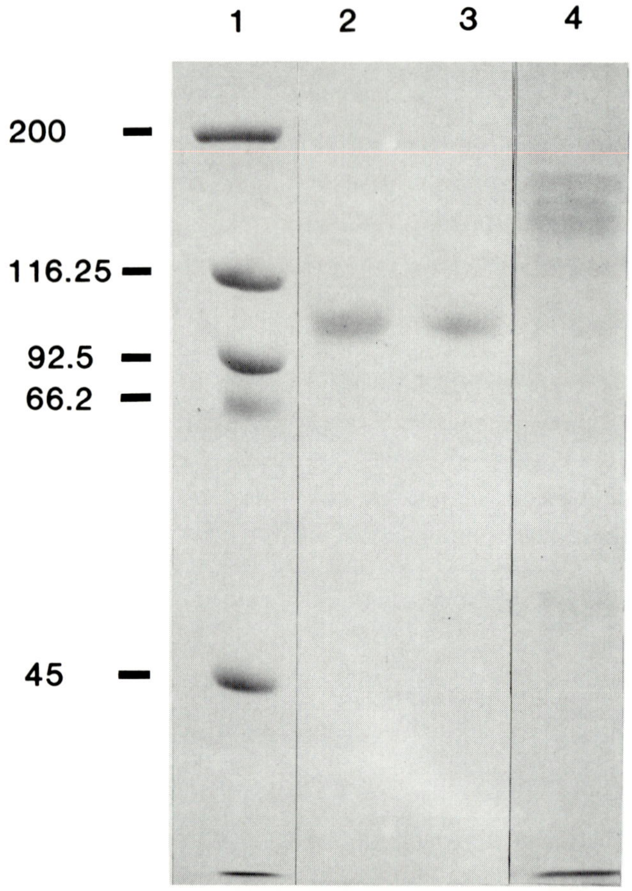

FIGURE 2. Analysis of proteins secreted by transformants containing the glucoamylase gene. Lane 1: molecular weight standards; lane 2: 1 μg purified glucoamylase; lane 3: 1 mℓ pPM16 culture supernatant; lane 4: 5 mℓ pMAC101 culture supernatant. The samples were analyzed on an SDS-urea polyacrylamide gel.

C. Fermentation of Soluble Starch by Recombinant Yeast Strains

Yeast transformants containing the glucoamylase gene were grown in liquid medium containing 10% soluble starch as a sole carbon source. These fermentations were carried out under air-restricted conditions. CO_2 production was estimated from weight loss measurements. Commercial glucoamylase was added as indicated in Figure 3 at approximately 1 U/mℓ (one glucoamylse unit was defined as the amount of enzyme that released 1 μmol of glucose per minute at pH 4.5 and 37°C). Under these conditions C468:pAC1 did not ferment starch, while C468:pGAC9 utilized 95% of the starch in the medium. Addition of commercial glucoamylase to the culture allowed growth of C468:pAC1 on soluble starch and enhanced the fermentation of C468:pGAC9 to a rate similar to that on glucose. It should be noted that the time course of starch fermentation is affected by the initial cell density. At higher inoculation levels, the starch fermentation is completed at earlier times.

D. Physiological Studies With Recombinant Strains

Table 1 lists the growth rates and fermentation rates of different yeast strains on glucose. No significant differences are observed between the parent and the cloned strains of C468 and C162. Expression and secretion of glucoamylase, up to 1.2% of the cell protein (strain C468pPM18), do not have a deleterious effect on yeast physiology.

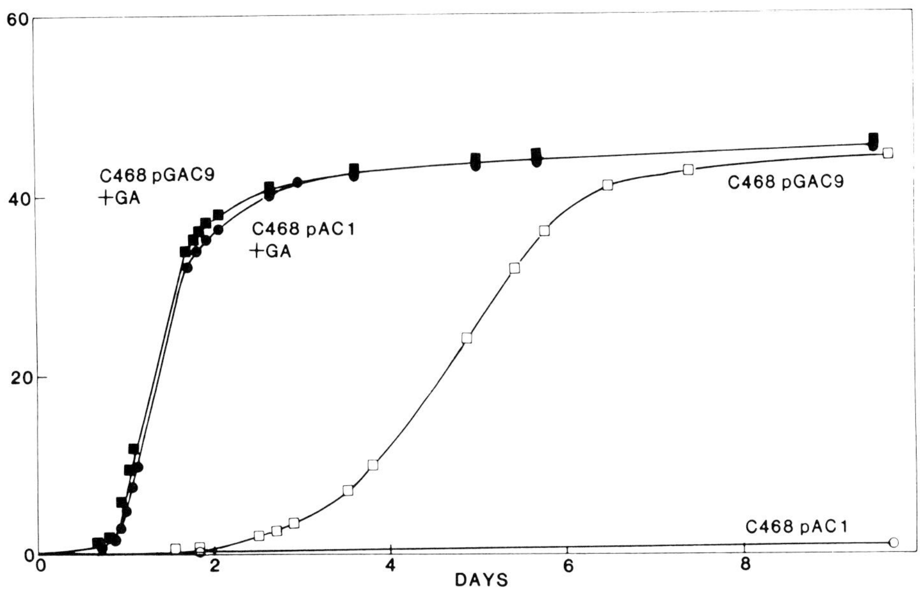

FIGURE 3. Fermentation of soluble starch by recombinant yeast strains with and without glucoamylase (GA). Fermentation of 10% (w/v) soluble starch was carried out under air-restricted conditions. CO_2 production was estimated from weight loss measurements. Commercial glucoamylase was added as indicated at approximately 1 U/mℓ. (Reprinted from Innis, M. A. et al., *Science*, 228(4695), 1985. With permission.)

Table 1
GROWTH AND FERMENTATION RATES OF RECOMBINANT YEAST STRAINS ON GLUCOSE (10% w/v)

Strain	Growth rate (h^{-1})	Fermentation rate (h^{-1})
C468	0.25	—
C468pGAC9	0.25	—
C468pPM18	0.23	—
C162	0.47	0.44
C162pPM20	0.48	0.44

The glucoamylase plasmid has been integrated into the chromosome of the distiller's strain (C162). Studies with this strain (C162pPM20) have shown no sign of glucoamylase loss during 800 generations of continuous growth under nonselective conditions. The maintenance of the glucoamylase gene for so many generations further shows that expression and secretion of glucoamylase in yeast have no deleterious effect on the yeast.

Table 2 lists the fermentation products of C162 and C162pPM20 on 10% glucose and 10% soluble starch. About 95% of the soluble starch was fermented. No difference in fermentation products was observed between the two strains on glucose. The expression of glucoamylase had no effect on cell or ethanol yield. On starch, fermentation by C162pPM20 is characterized by lower glycerol and higher ethanol yields than comparable fermentation on glucose media. The lower glycerol production on starch is probably due to the lower osmotic pressure in the starch medium. The higher ethanol yield is presumably due to lower glycerol production.

Table 2
FERMENTATION PRODUCTS OF C162 AND C162pPM20 ON GLUCOSE AND SOLUBLE STARCH

	Carbohydrates utilized (g/ℓ)	Cells (g/ℓ)	Ethanol (g/ℓ)	CO_2 (g/ℓ)	Glycerol (g/ℓ)	Ethanol yield (g/g)	Carbon recovery (%)
10% Glucose							
C162	99.0	5.61	44.7	42.0	6.13	0.45	99
C162pPM20	99.0	5.45	44.5	41.8	6.47	0.45	99
10% Soluble starch							
C162	5.0	2.44	—	3.5	0.84	—	101
C162pPM20	93.0	7.07	44.8	43.0	1.62	0.48	104

III. SUMMARY

- We cloned, characterized, and expressed the *A. awamori* glucoamylase in yeast.
- We integrated the glucoamylase gene into the genome of an industrial yeast.
- The glucoamylase is secreted by yeast at levels sufficient to allow growth on liquified starch.
- The secreted glucoamylase is similar to the native enzyme.
- In medium containing liquified starch 93 to 95% of the starch was fermented by laboratory and industrial strains containing the glucoamylase gene.
- Fermentations of an industrial distiller's strain containing an integrated glucoamylase gene had higher ethanol and cell yields, and lower glycerol yield.
- There was no loss of glucoamylase activity during 800 generations of continuous growth under nonselective conditions.

REFERENCES

1. **Nunberg, J. H. et al.,** *Mol. Cell. Biol.,* 4, 2306, 1984.
2. **Innis, M. A. et al.,** *Science,* 228, 21, 1985.
3. **Boel, E. et al.,** *EMBO J.,* 3, 1097, 1984.

Chapter 13

THE YEAST KILLER TOXIN EXPRESSION SYSTEM: FUNCTIONAL ANALYSIS AND BIOTECHNOLOGICAL APPLICATIONS

David Y. Thomas, Thierry Vernet, Charles Boone, Deirdre Greene, Susan Lolle, and Howard Bussey

TABLE OF CONTENTS

I. INTRODUCTION

The ability of cells to produce toxins is widespread in nature. For many of these toxins, from a variety of species, there is a remarkably similar structure. They are often composed of two polypeptides which have different functions. One polypeptide is associated with lethality, the other with binding to the target cell.[1] Some yeast strains produce a bipartite toxin which can kill other yeasts.[2,3] We have been examining the organization and function of the *K1* toxin using in vitro genetic manipulation. The yeast toxin is different from other eukaryotic toxins in that it is specific for other cells of the same species. The producing cells are themselves immune to the action of the toxin, similar to the *Escherichia coli* colicins. The *Saccharomyces cerevisiae* toxin is coded for by a small double-stranded RNA termed *M1*. The recent cloning of the portion of this dsRNA containing the toxin gene into cDNA[4,5] and its expression in yeast[6] has enabled us to define the function of the different parts of the molecule using site-directed mutagenesis.[7]

Among the proteins that *S. cerevisiae* cells secrete, the α-hormone[8] and yeast-killer toxin[2,3] are synthesized as intracellular precursors which are proteolytically processed in a manner reminiscent of some mammalian peptide hormones, plant storage proteins, and some other toxins. The processing pathway and the processing enzymes can be readily manipulated in the yeast system and some interesting results for the processing of polyproteins, in general, have been obtained from the yeast system.

Here we review our current knowledge of the structure of the toxin molecule, its processing from the precursor preprotoxin, and our understanding of its mode of action. Additionally, we review some applications of the yeast-killer system to biotechnological processes.

II. STRUCTURE OF THE KILLER TOXIN PRECURSOR

The double-stranded RNA coding for the yeast-killer toxin has been cloned into a cDNA copy and sequenced.[4-6] An open reading frame was identified that coded for a protein the size of the intracellular precursor previously identified.[6] From the protein sequences of the subunits of the extracellular toxin, the positions of these subunits within the precursor were established. The carboxyl termini of the subunits have not been unequivocally determined, but based on the molecular weight of the subunits the α subunit probably terminates at *tryptophan* residue *130* and the β subunit probably extends to the end of the precursor, *histidine 316* (Figure 1). The proteolytic enzymes believed to cleave at sequences are defined by two mutations *kex1* and *kex2*, both of which result in a killer-defective phenotype.[2,3]

The overall secretion pathway of the toxin (both that coded by the M1 dsRNA and the cDNA expression vector) follows that of other secreted proteins in yeast, which in turn is similar to the organization of the secretory apparatus in other eukaryotic cells.[6,9] The progression of the yeast toxin precursor (preprotoxin) through this pathway is accompanied by at least three proteolytic cleavage events. Genetic evidence is that the *KEX2* cleavage of the precursor occurs as a late Golgi/secretory vesicle event and is preceded by the *KEX1* cleavage process.

Entry into the secretory pathway is facilitated by a hydrophobic leader sequence, which is atypical of secreted proteins in yeast in that it is 44 amino acids long. There is little definitive evidence of whether the cleavage of this leader occurs cotranslationally or post-translationally. It is not certain if the N terminus of the α subunit is the leader peptidase cleavage site or whether there is yet another cleavage site in this leader.

Using an open-reading-frame yeast expression vector[6] we have created a deletion which completely removes the leader peptide and alters the extreme N terminus of the α-subunit. This construct secretes a small amount of active toxin which is capable of killing yeast spheroplasts, but not whole cells. In addition, cells carrying this plasmid are partially immune

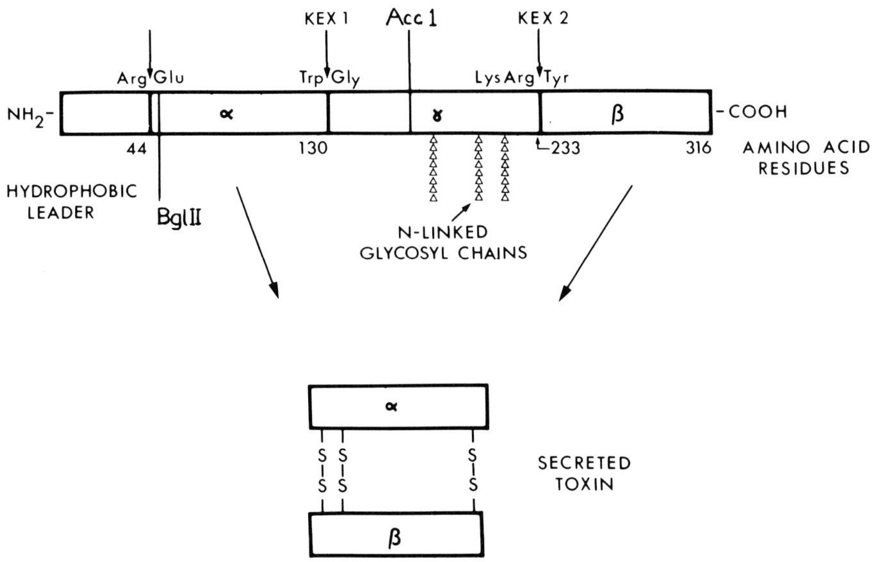

FIGURE 1. The structure of the preprotoxin molecule is shown at the top of the figure. The molecular weight determined from the DNA sequence (34.7 kdaltons) corresponds well in size to the nonglycosylated toxin. The peptidase cleavage site after the 44 amino acid leader sequence is indicated, as are the KEX1 and KEX2 cleavage sites. The structure of the extracellular α/β toxin dimer is shown below. The location of the disulfide bonds between the two subunits has not yet been determined.

to toxin, and a nonglycosylated intracellular precursor can be detected.[9] How does such a molecule enter the secretory pathway and get processed? There are several possibilities: one is that one or both of the hydrophobic domains in the α-subunit adopt a conformation which allows insertion of the molecule into the endoplasmic reticulum. A clear understanding of how this leaderless mutant functions awaits more data.

III. ASSIGNMENT OF FUNCTIONS TO THE PREPROTOXIN

It has proved difficult to separate in vitro the α and β subunits of the yeast-killer toxin and to assay their functions. The toxin has a limited half-life even under conditions which stabilize the toxin, and recombination of the subunits has not yet been possible. We have employed a different approach to assign functions to the different regions of the preprotoxin.[7] We have constructed a series of mutations throughout the preprotoxin molecule; these mutations were constructed in vitro using three methods:

1. N- and C-terminal deletions to various specific restriction enzyme sites within the gene
2. Insertion of amino acid sequences designed to disrupt protein structure by insertion of oligonucleotides at restriction enzyme sites
3. Alteration of one or insertion of two amino acids by oligonucleotide-primed single-stranded site-directed mutagenesis

There are two easily assayed functions associated with the killer toxin, the ability to kill sensitive yeast cells and the ability to confer immunity. We have used these phenotypes to assess the effects of the constructed mutants. We have also used antibodies to detect intracellular preprotoxin. The glycosylation of this molecule has served as an indicator of its entry into the secretory pathway.

IV. LOCATION OF THE IMMUNITY DOMAIN

It has been inferred for some time that immunity is coded for by the *M1* dsRNA.[2,3] The phenotype is readily assayed; cells become resistant to added toxin. The cDNA clone which we constructed contains only the region of the *M1* dsRNA molecule coding for the preprotoxin, and confers both immunity and the ability to produce toxin. Apart from the reading frame which codes for a polypeptide, the size of the preprotoxin, there is no other extended open reading frame on either DNA strand. An initial and key observation was that *kex1 kex2* double mutants are immune.[3,11] They produce greatly reduced levels of toxin indicating that little processing of the preprotoxin occurs. Thus, it appears that the preprotoxin molecule itself could act to confer immunity.

We have used site-directed mutagenesis to determine in which region of the preprotoxin the immunity domain is located. It had been suggested that the γ region of the preprotoxin was responsible for expressing the immunity phenotype.[3,11] Therefore, we initially constructed mutants in the γ region. To summarize the results we have so far obtained — none of the mutations which have altered a substantial part of the γ region, or which completely deleted the β region, abolished immunity. However, various mutations in the α region perturb both the killer and immunity phenotype. Interestingly, the mutations in the γ region prevented or substantially reduced secretion of active toxin, but all contained glycosylated preprotoxin molecules of appropriate size. This reinforces the notion that the preprotoxin itself can confer immunity.

We looked for free γ protein within cells using antibodies raised to synthetic peptides from within the γ region. Using this approach we were unable to detect free γ peptide within or secreted from cells, even though the antibody precipitated the preprotoxin from in vitro translates. Thus, the free γ peptide does not appear to be involved directly in immunity, neither is it stably maintained in the cell after its processing from the precursor. However, it may play an essential role in the processing of the precursor, as mutations we have made in this region appear to profoundly affect proteolytic processing of the precursor. These mutant precursors do enter the secretory pathway as they become glycosylated. The simplest interpretation of this effect is that the action of the *KEX1* and *KEX2* encoded protease activities are sensitive to the conformation of the substrate.

The overall structure of the preprotoxin molecule is similar to that of preproinsulin. The C chain of preproinsulin is thought to be concerned with the folding of the disulfide-linked B and A subunits. However, the C chain shows a high degree of diversity between different species, indicating that a variety of sequences can accommodate this structural requirement. It will be of wide interest to determine how the mutations in the interstitial γ region affect processing at the remote proteolytic cleavage sites. Also, we plan to search for second-site mutations in the precursor which can correct processing of some of these mutants.[23] Even so, does the γ region code for a measurable function? In mammalian cells precursors of neuropeptides often contain more than one hormone. In the gene for the precursor polypeptide of yeast α-hormone there is a mutation which leads to an altered response of the host cell to added hormone.[24] The lack of stability of the free γ peptide argues against such a role, or that any exertion of its effects depends on its transitory appearance.

There are two hydrophobic regions in the α subunit separated by a charged domain which are potentially membrane inserting. In-frame insertion of two amino acids in these hydrophobic regions gave a killer/immunity-defective phenotype. An explanation for the simultaneous loss of toxin production and the immunity phenotypes would be simply that the precursor fails to enter the secretory pathway. Yet, in both mutants the glycosylated preprotoxin was detected, thus the mutant precursors enter the secretory pathway. The possibility that the mutations prevent the precursor from reaching the location where it generates immunity rather than affecting the immunity domain per se has been ruled out, since one of the mutants secretes a detectable, but inactive, toxin.

In conclusion, our results suggest that the α subunit of the preprotoxin is involved in the expression of the immunity phenotype. Similarly, the integrity of the γ region is probably required for correct processing of the preprotoxin at the *kex1* and *kex2* sites.

V. MODELS OF IMMUNITY

The models we have of both toxin and immunity action rely heavily upon the concept of a specific toxin membrane receptor (see later). It is clear that the α region of the preprotoxin is somehow involved in expression of immunity. We suggest that the α region of the preprotoxin binds to the toxin receptor, thus preventing further lethal interaction with active toxin. Hence, this model implies that the receptor-bound preprotoxin is stable and acts as a competitive inhibitor of active toxin.

VI. PROTEOLYTIC PROCESSING OF THE PREPROTOXIN

The proteolytic processing of the precursor is defined by two genes, *KEX1* and *KEX2*. There is evidence that KEX1 acts before *KEX2* in the processing of the preprotoxin, in that in a double mutant *(kex1 kex2)* the precursor has the phenotype of a single mutant (*kex1*),[11] and it is unprocessed. The *KEX2* activity is required for the processing of both killer and α-hormone precursor proteins. The gene coding for *KEX2* has recently been cloned[12] and sequenced and shown to be the structural gene for a calcium-requiring protease which has a thiol in its active site. The enzyme displays a preference for cleaving substrates which contain at least one arginine and is particularly active on *arg-arg* and *lys-arg* residues. The enzyme appears to be membrane bound and its cellular location is the secretory granules and, presumably, the processing of precursors takes place within this organelle. The *KEX2*-encoded protease activity is also required by other secreted proteins of yeast, as examination of the extracellular proteins of *kex2* mutants shows many altered proteins.[13] There is good evidence from mutants of the role of the *KEX2* protease in the processing of α-hormone dodecamers from the precursor. A *kex2* mutant does not secrete active α-hormone or killer toxin. We plan to alter the sequence around the amino acid residues at the *lys-arg-tyr KEX2* cleavage site in the preprotoxin, to examine which other residues and structural features target the cleavage into this site rather than dibasic residues elsewhere in the precursor.

The *KEX1* gene is known to be located on chromosome *VII* of yeast.[2,3] Mutations in this gene have the phenotype that very little active toxin is secreted in common with *KEX2* mutants. The defined role of the *KEX2* calpain-type protein in the processing of yeast-killer toxin and α hormone is established and indicates the role of the calpains in other species. The question arises if similarly there is evidence for *KEX1*-type cleavage sites in protein precursors in other species. Potential candidates for chymotrypsin-type cleavage sites are present in the sequences of the precursor proteins for the plant toxin ricin[14] and for atrian-aturetic factor.[15] The poliovirus polyprotein, which is extensively processed by a viral encoded protease, is also processed by a host chymotrypsin-like protease at two sites which have *tyr-gly* pairs.[22]

VII. THE TOXIN RECEPTOR

Most toxins kill another cell type. In principle, the only immunity requirement for such toxins would be that there be no similar receptor on the producing cells. It has been postulated, although the analogy may not be a correct one, that some toxins owe their action to being similar to polypeptide hormones and use the receptor for that hormone.[1]

What is the evidence for a receptor for the killer toxin? For yeast α hormone there is good genetic and kinetic evidence for the existence of a receptor for both α[16] and a-hormones.[17] These receptors are defined by mutations which are insensitive to each of these hormones. There is evidence that toxin binds specifically to a (1→6)-β-D-glucan present in the yeast wall.[18] The mutants *kre1* and *kre2* were selected as being resistant to killer toxin; in these mutants radioactive toxin shows a reduced binding to the cell wall. The binding sites on the cell wall are of lower affinity than those on the cell membrane, as at concentrations at which the wall binding sites are not saturated, cell killing is still observed. That is, killing is achieved at lower concentrations than those required to saturate the wall binding sites, implicating other higher affinity sites. That the location of these is probably on the cell membrane is indicated by the fact that spheroplasts can be killed by toxin, even those prepared from the *kre1*- and *kre2*-resistant mutants. These observations are consistent with a model whereby the receptor is located on the cell membrane and the cell wall receptors have a lower affinity for the toxin. There is no information on the identity of the membrane receptor, however, there is a mutation *kre3* which has the normal cell wall binding sites, but the spheroplasts are partially resistant to toxin.[18] This mutation also confers resistance to similar toxins from a variety of other yeast genera. This is the phenotype expected of a mutation in the cell membrane receptor. The cloning and identity of this receptor are awaited with interest.

What is the natural role of this receptor? It is obviously not primarily present in the yeast cell wall to bind killer toxin. The mechanism of low affinity receptors located on the cell wall and higher affinity receptors on the cell membrane might be used naturally by yeast for metabolites or signals which are present at low concentrations in the environment, and this role has been usurped by the toxin. The receptors do not appear to be the same as the receptors for α and a-hormones. These are defined by the hormone-insensitive mutants *ste2* and *ste3*, respectively.[16,17] These mutants are still sensitive to toxin and probably do not code for the toxin receptor.

VIII. MODELS OF TOXIN ACTION

The problem of how the interaction of a toxin with the plasma membrane of a cell leads ultimately to a lethal event has not been solved in detail for any toxin.[1] There are a number of toxins which have adopted a similar overall strategy for cell killing. Like the yeast-killer toxin they have two polypeptide components held together by disulfide bonds. In the case of ricin the β subunit binds to the cell surface receptor and somehow facilitates the entry of the α subunit.[1]

There is good evidence that yeast-killer toxin interacts with the plasma membrane in an energy-dependent manner. Toxin-treated cells leak potassium and this may be the cause of cell death.[18] A study with an analogous toxin from the yeast *Pichia* has shown that this toxin can insert into artificial lipid bilayers and form conductance channels.[19] Recent experiments have shown that toxin exists in the medium as multimers of four to eight α-β dimers.[9] Such aggregates, if they are the active form of the toxin, could associate so that the α helical regions of the α subunits form an ion channel in membranes.

IX. APPLICATIONS OF THE KILLER TOXIN SYSTEM TO BIOTECHNOLOGY

There are several applications of the killer toxin system which are of interest in biotechnological processes and their development.

The first applications concern the phenotype conferred on cells by the killer toxin. This is the ability of colonies of cells carrying the killer toxin gene to form killing zones on a lawn of sensitive cells. This easily scored phenotype has been used for a variety of purposes. The *ski5* host mutation was isolated as a mutant having a larger zone of killing.[20] Subsequent analyses have shown that the host cells are deficient in an extracellular protease, which appears to be distinct from the α-hormone-degrading *bar1* phenotype. Toxin and some other secreted yeast proteins are degraded by this protease and strains defective in this activity are of interest in the extracellular production of heterologous proteins by yeast.

Another application arises from the observation that the cDNA plasmid with the toxin gene transcribed from the ADH1 promoter produces more abundant transcripts than the dsRNA killer, yet the amount of secreted toxin is the same. Presumably, this is due to some limitations of the flux of toxin through the secretory pathway. We have mutagenized host cells carrying this plasmid and screened them for killing zones of increased diameter. We are now examining these mutants to see if they are specific for killer or act, in general, on secreted proteins. This strategy can be employed to increase the amounts of heterologous proteins secreted by yeast.

The screening system can also be employed for looking at host mutants altered in protein transport and also for plasmid mutations affecting the same process. A method we are using for looking at mutants generated by site-directed mutagenesis in specific regions of the toxin precursor is to look for suppressor mutations in the precursor which restore killer function.[23] In the case of preprotoxin mutant precursors which cannot be processed (e.g., in the γ region), a search was made for host mutations which can correctly process the mutant precursor into active toxin. Such mutants may be expected to involve other components of the proteolytic processing system.

The *KEX1* and *KEX2* protease activities associated with the processing of the preprotoxin are of interest because of their specificity and the opportunity to characterize at a molecular level two proteins involved in the processing of polyproteins. The nature of the interactions of these enzymes with their natural substrate in the secretory pathway is accessible through site-directed mutagenesis of the proteolytic processing sites of the preprotoxin, and through mutagenesis of the *KEX1* and *KEX2* genes.

In more direct applications to yeast biotechnology we have used the killer cDNA expression plasmids to develop a transformation system. The fact that killing and immunity reside on the same molecule has been used in this application. There are only a few dominant selection systems available for the selection of yeast transformants. Using the killer expression plasmids it is possible to transform yeast hosts which have no selectable genetic markers.[9,21] This is particularly of interest for the transformation of industrial yeast strains which are often polyploid. Nontransformed cells have been eliminated by treatment with toxin. We have found that the incorporation of methylene blue into plates buffered at pH 4.7 allows transformants to be simply detected as blue pocks in a confluent background of sensitive, nontransformed cells.[9] This nonselective screening system is very sensitive in detecting rare transformants in a variety of yeasts used in different industrial processes. This system can be used to introduce other genes into polyploid yeasts by construction of the appropriate plasmid. For example, we have incorporated the gene for a yeast glucoamylase into the killer expression plasmid and used this to transform industrial yeast strains. These transformants now secrete the glucoamylase and the toxin.[9] Further modifications of these plasmids to remove the *Escherichia coli* plasmid segments will make the constructions entirely of yeast origin, which may be required for some applications in the food industry. This toxin-based transformation system provides an entry into the molecular genetics of industrial yeast strains.

For multicopy plasmids in yeast, as in other organisms, plasmid loss under nonselective conditions is significant. An autoselection procedure based upon the killer expression plasmids has been devised.[21] The extracellular toxin is most stable at pH 4 to 5 and at temperatures of 20 to 25°C, conditions under which many industrial fermentations are performed. In a brewing strain of yeast, the killer expression plasmid was stably maintained in yeast complex media and in wort medium at pH 4 and 20°C. At pH 7 the plasmid was rapidly lost, and if cultures in which less than 10% of the cells possessed the plasmid were adjusted to pH 4.5, the cultures rapidly returned to a condition where all the viable cells contained the plasmid.[21] This simple system appears to have some application in the employment of genetically engineered yeasts in industrial processes.

Utilization of the toxin leader sequence to secrete heterologous proteins from yeast is another obvious use of this system. However, it is not clear if the toxin leader has any advantages over the leaders of other secreted yeast proteins in this application. The proteolytic processing sites, either at *KEX1* or *KEX2*, do have some interest for the processing and secretion of heterologous proteins by yeast. Such proteins may be correctly processed to have the authentic amino terminus. The exact role of glycosylation in the secretion of proteins is not clear, but it does appear that the glycosylated precursor is more efficiently secreted. The *KEX1* and *KEX2* sites provide a means where the glycosylated portion of a protein can be separated from the final secreted peptide.

X. CONCLUSION

The killer system, now that it is genetically accessible through the cDNA expression clones, has some interesting applications. As an example of a disulfide-linked toxin, it has the advantage that, unlike, for instance, ricin, the effects of site-directed mutations can be directly assayed. The components involved in this toxin system are all potentially clonable and experimentally manipulable. There are also some applications to industrial yeasts which will be useful for further development of the genetics of these strains.

Note Added in Proof

Recent results have added considerably to our knowledge of how the yeast killer preprotoxin molecule is processed and to the location of the immunity domain. By means of a series of site-directed mutations the immunity domain was found to be located at the C terminus of the α-subunit and the N terminus of the γ-subunit (Figure 1).[25] The location of the *KEX1* cleavage site was subsequently shown to be not at $tryptophan_{130}$, but that the α-subunit extends to $alanine_{147}$. It is possible therefore that the immunity domain is totally contained within the α-subunit.[9] The *KEX1* gene has been cloned and sequenced. The gene has, in part, a high degree of homology with the sequence of yeast carboxypeptidase Y.[9] A possible mode of processing of the preprotoxin precursor would therefore be a cleavage by the endoproteinase *KEX2* at $arginine_{149}$, followed by the removal of $arginines_{149}$ and $_{148}$ by the carboxypeptidase action of *KEX1*.

ACKNOWLEDGMENTS

We thank M. Ahmad, H. Zhu, and L. Hougan for assistance and helpful discussion in parts of this work. We also thank D. Harrison and Dr. R. Brousseau for synthesis of oligonucleotides and B. Soldner for excellent technical assistance. This work was supported in part by operating and strategic grants from the Natural Sciences and Engineering Research Council of Canada, issued as NRCC #25035.

REFERENCES

1. **van Heynigen, S.,** Similarities in the actions of different toxins, in *Molecular Action of Toxins and Viruses,* Cohen, P. and van Heynigen, S., Eds., Elsevier, Amsterdam, 1982, 169.
2. **Bussey, H.,** Physiology of killer factor, *Adv. Microb. Physiol.,* 22, 93, 1981.

3. **Tipper, D. J. and Bostian, K.,** Double-stranded ribonucleic acid killer systems in yeast, *Microbiol. Rev.*, 48, 125, 1984.
4. **Skipper, N., Lau, P., and Thomas, D. Y.,** Cloning and sequencing of the preprotoxin-coding region of the yeast M1 double-stranded RNA, *EMBO J.*, 3, 107, 1984.
5. **Bostian, K. A., Bussey, H., Elliot, Q., Burn, V., Smith, A., and Tipper, D. J.,** Sequence of the preprotoxin gene of type 1 killer yeast: multiple processing events produce a two component toxin, *Cell*, 36, 741, 1984.
6. **Bussey, H., Saville, D., Greene, D., Tipper, D. J., and Bostian, K. A.,** Secretion of yeast killer toxin: processing of the glycosylated precursor, *Mol. Cell. Biol.*, 3, 1362, 1983.
7. **Vernet, T., Boone, C., Greene, D., Thomas, D. Y., and Bussey, H.,** Yeast killer toxin: site-directed mutations affecting toxin and immunity domains of the precursor gene, in *Natural Antimicrobial Systems*, Bath, U. K. and Board, R. G., Ed., Bath University Press, in press.
8. **Julius, D., Blair, L., Brake, A., Sprague, G., and Thorner, J.,** Yeast alpha-factor is processed from a larger precursor polypeptide: the essential role of a membrane-bound dipeptidyl aminopeptidase, *Cell*, 32, 839, 1983.
9. Unpublished results.
10. **Hutchins, K. and Bussey, H.,** Cell-wall receptor for yeast killer factor: involvement of a (1→6)-beta-D-glucan, *J. Bacteriol.*, 154, 161, 1983.
11. **Bussey, H.,** Yeast killer toxin and an immunity gene are processed from a composite precursor, *Microb. Sci.*, 1, 62, 1984.
12. **Fuller, R. S., Brake, A. J., Julius, D. J., and Thorner, J.,** The KEX2 gene product required for processing of yeast prepro-alpha-factor is a calpain-like endopeptidase specific for cleaving at pairs of basic residues, in *Protein Transport and Secretion*, Gething, M. J., Ed., Cold Spring Harbor Laboratory, Cold Spring Harbor, N.Y., 1985, 97.
13. **Rogers, D. T., Saville, D., and Bussey, H.,** *Saccharomyces cerevisiae* killer expression mutant *kex2* has altered secretory proteins and glycoproteins, *Biochem. Biophys. Res. Commun.*, 90, 187, 1979.
14. **Lamb, F. I., Roberts, L. M., and Lord, J. M.,** Nucleotide sequence of cloned cDNA coding for preproricin, *Eur. J. Biochem.*, 148, 265, 1985.
15. **Thibault, G., Lazure, C., Schiffrin, E. L., Gutowska, J., Chartier, L., Garcia, R., Seidah, N. G., Chretien, M., Genest, J., and Cantin, M.,** Identification of a biologically active circulating form of rat atrialnatriuetic factor, *Biochem. Biophys. Res. Commun.*, 130, 981, 1985.
16. **Jenness, D. D., Burkholder, A. C., and Hartwell, L. H.,** Binding of alpha-factor pheromone to yeast *a* cells: chemical and genetic evidence for an alpha-factor receptor, *Cell*, 35, 678, 1983.
17. **Herskowitz, I.,** Yeast as the universal cell, *Nature*, 316, 678, 1985.
18. **Skipper, N. and Bussey, H.,** Mode of action of yeast toxins: energy requirements for *Saccharomyces cerevisiae* killer toxin, *J. Bacteriol.*, 129, 668, 1979.
19. **Kagan, B. L.,** Mode of action of yeast killer toxins: channel formation in lipid bilayer membranes, *Nature*, 302, 709, 1983.
20. **Bussey, H., Steinmetz, O., and Saville, D.,** Protein secretion in yeast: two chromosomal mutants that over-secrete killer toxin in *Saccharomyces cerevisiae*, *Curr. Genet.*, 7, 449, 1983.
21. **Bussey, H. and Meaden, P.,** Selection and stability of yeast transformants expressing cDNA at an M1 killer toxin-immunity gene, *Curr. Genet.*, 9, 285, 1985.
22. **Pallansch, M. A., Kew, O. M., Semler, B. L., Omilianowski, D. R., Anderson, C. W., Wimmer, E., and Rueckert, R. R.,** Protein processing map of poliovirus, *J. Virol.*, 49, 873, 1984.
23. **Shortle, D. and Lin, B.,** Genetic analysis of staphylococcal nuclease: identification of three intragenic "global" suppressors of nuclease-minus mutations, *Genetics*, 110, 539, 1985.
24. **Kurjan, J.,** Alpha-factor structural gene mutations in *Saccharomyces cerevisiae*: effects on alpha-factor production and mating, *Mol. Cell. Biol.*, 5, 787, 1985.
25. **Boone et al.,** *Cell*, 46, 105, 1986.

Chapter 14

EXPRESSION OF HUMAN TISSUE PLASMINOGEN ACTIVATOR (t-PA) IN *SACCHAROMYCES CEREVISIAE*

G.S. Gill, P.G. Zaworski, K.R. Marotti, L.M. Bennett, and D.R. Thomsen

TABLE OF CONTENTS

I. ABSTRACT

t-PA, a 64,000-dalton secreted glycoprotein, is a serine protease that plays an important role in blood clot dissolution. It activates plasminogen to plasmin which degrades fibrin in the clot. In order to see whether yeast could synthesize a complex mammalian protein like t-PA, we cloned the t-PA cDNA in a yeast expression vector behind the MFα1 promoter and leader sequences. This vector construction directed the synthesis of active t-PA. Maximal amounts of t-PA activity were detected during the logarithmic growth phase. There was a rapid drop in t-PA activity when cells entered the stationary phase. Very little activity was found in the culture fluid. Cells from a culture expressing t-PA when plated on minimal medium supplemented with casamino acids formed large and small colonies. The small colonies did not grow on absolute minimum medium and had no t-PA activity. Also, cells grown in the minimum medium showed maximum t-PA activity.

Partially purified t-PA from yeast cells, when run on an activity gel, gave a clear zone at the same place as the t-PA purified from a Bowes melanoma cell line, implying glycosylation of t-PA expressed in yeast. An endoglycosidase (Endo H) treatment of the ^{14}C-labeled yeast t-PA indicated such was the case. Also, Endo H treatment of extracts made from cells expressing t-PA resulted in a threefold increase in t-PA activity. It is likely yeast is hyperglycosylating some of the t-PA which becomes inactive.

II. INTRODUCTION

t-PA is an enzyme of considerable therapeutic importance. It belongs to a class of proteases which are involved in dissolving blood clots. These proteases are thought to mediate their effect through activation of plasminogen to plasmin. Plasmin, in turn, degrades fibrin networks in the clot leading to the dissolution of the clot. Unlike other plasminogen activators, e.g., urokinase and streptokinase, t-PA activates plasminogen mainly at the site of the clot and thus does not stimulate generalized systemic fibrinolysis.[1,2]

Cloning of cDNA copy of t-PA gene has been reported.[3,4] The deduced amino acid sequence shows that it is a secreted glycoprotein with four N-linked glycosylation sites. The native molecule has 35 cysteine residues with a potential for the formation of 17 intramolecular disulfide bridges in the folded protein.[3] Such a complex molecule, if expressed cytoplasmically, would, most likely, be unable to fold properly, for the cytoplasmic environment is too reducing. Indeed, high level cytoplasmic expression of t-PA in *Escherichia coli* produced totally inactive protein.[3,5] Activation through in vitro refolding of such a complex protein would be a formidable task. In order to circumvent these problems we targeted t-PA to the yeast secretory pathway by hooking the gene to the signal sequences from the yeast MFα1 gene. MFα1 codes for one of the yeast mating pheromones, the α factor, and has been described in detail.[6,7] The signal sequences have been shown to direct heterologous secreted proteins to the yeast endoplasmic reticulum and beyond to the periplasmic space, and, in some cases, to the outside of the cell.[8,9]

In this report we show that introduction of t-PA gene fused to the MFα1 promoter and signal sequences into yeast cells results in the synthesis of active t-PA. The activity is plasminogen dependent and has a molecular weight similar to that of the activity purified from Bowes melanoma cells. Treatment of "yeast" t-PA with Endo H indicates that the material is glycosylated.

III. MATERIALS AND METHODS

A. Media and Strains

Standard yeast and bacterial culture media were used for growth of the respective organisms.[10,11] YEPD consisted of 2% dextrose, 2% peptone, and 1% yeast extract. Yeast minimal

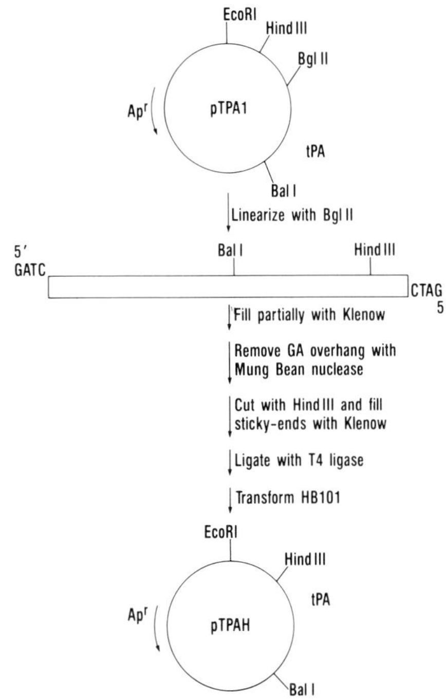

A

FIGURE 1. Construction of the expression vector pα1-TPA. The flow charts shown in (A) and (B) outline the steps involved in the construction of the shuttle vector used to express t-PA in yeast. In (A) the modifications made to the t-PA gene resulted in the removal of the "pre-pro" sequences of the gene.[3] In (B) we started with a shuttle vector carrying the promoter and "pre-pro" sequences of the yeast MFα1 gene.[6,7] The three-piece ligation gave rise to the vector pα1-TPA. This plasmid was used to transform the yeast strain YNN27.[12]

medium (YMM) was 2% glucose, 0.7% yeast nitrogen base (DIFCO). It was supplemented, if required, with other amino acids or uracil at a concentration of 50 to 100 μg/mℓ. YMMCA was YMM supplemented with 1% casamino acids. For solid medium, agar was added to the various media to 2%. Bacterial cultures were grown in L broth. The yeast strain YNN27 (α, ura3-52, trpl-289, gal2), obtained from R. Davis, was used as a host for t-PA expression, while HB101 was the bacterial host employed for all the vector constructions. All the restriction enzymes, T4 ligase and DNA polymerase I (Klenow fragment), were purchased from New England Biolabs and used as prescribed by the manufacturer.

B. Construction of t-PA Expression Vector

The flow charts shown in Figure 1 give details of the construction of the vector pα1-TPA. The t-PA gene was inserted in the expression vector pα1-ADHt in such a way that the MFα1 "pre-pro" sequences and the t-PA were in the same reading frame. This will ensure expression and correct processing of the hybrid gene product in yeast. Further details are given in the legend to Figure 1.

C. t-PA Activity Assay

t-PA activity was assayed using a modification of the method of Tang et al.[13] The following assay components were maintained as stocks: plasminogen stock solution (Kabi Vitrum, 1 casein unit [cu]/mℓ stored at −20°C), chromogenic substrate H-D-valyl-L-leucyl-L-lysine-

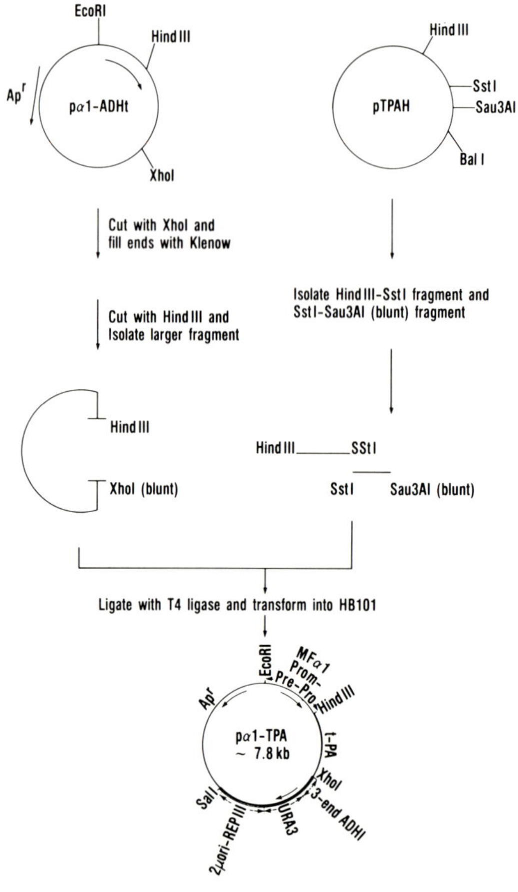

FIGURE 1B

p-nitroanilide dihydrochloride (Kabi Vitrum, 3-2251, 3.02 m*M* stored at 4°C), heparin (20 U/mℓ, stored at 4°C), 0.1 *M* Tris-HCl, 0.1% Triton® X-100, pH 8.0 (assay buffer). YNN27 cells with or without the plasmid pα1-TPA were grown at 30°C with aeration in YMM containing 100 μg/mℓ tryptophan, harvested by a slow-speed spin in a centrifuge, and resuspended in 0.5 to 1.0 mℓ of assay buffer. The cells were disrupted by vortexing with 0.5-mm-diameter glass beads using a Biospec® minibead beater. Cell lysates were removed with a pipette and cleared in an Eppendorf® minifuge with a 20-sec spin. The cleared lysates were diluted 1:10 in assay buffer. t-PA activity was monitored in a 2-mℓ reaction mixture which consisted of 1.6 mℓ of assay buffer, 0.1 mℓ cleared lysate, 0.1 mℓ of 1-CU/mℓ plasminogen, 0.1 mℓ chromogenic substrate, and 0.1 mℓ 20-U/mℓ heparin. The reaction was carried out at 37°C and the change in absorbance per minute at 405 nm was measured using a Shimadzu recording spectrophotometer. In order to determine concentration of t-PA in the unknown samples, standard curves were generated by plotting reaction rates vs. known concentrations of t-PA. Such a plot is shown in Figure 2.

D. Endoglycosidase H Treatment of Cell Lysates

Cell lysates were prepared as described previously, but with the following modification. Cells were disrupted in 0.1 *M* citrate buffer, pH 5.5, containing 0.1% Triton® X-100. The lysates were incubated with and without Endo H (0.8 μg/mℓ) at 37°C. Samples were removed and diluted in assay buffer and assayed for t-PA activity. Protein concentrations were determined using BioRad® protein assay. Results are given in Table 1.

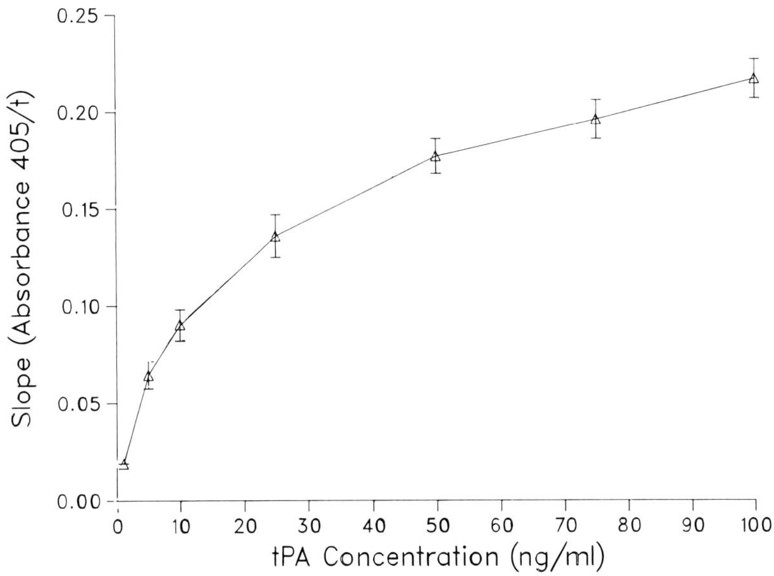

FIGURE 2. Construction of a standard curve for measurements of active t-PA in cell extracts. This curve was generated by measuring the slopes of curves in which change in absorbance as a function of time was measured for known concentrations of t-PA purified from Bowes melanoma cells. Plotting the slopes vs. concentration of t-PA gave us the standard curve. Thus, measuring the slope of the curve using an unknown sample would allow an estimate of the amount of active t-PA in the sample.

Table 1
RESULTS OF TREATMENT OF CELL LYSATES FROM *S. CEREVISIAE* YNN27 CONTAINING THE PLASMID Pα1-TPA WITH ENDO H

Incubation time (hr)	t-PA ng/mg protein		Relative activity increase
	+ Endo H	− Endo H	(+ Endo H/ − Endo H)
0	—	39.27	—
0.5	121.73	37.36	3.25
1.0	87.06	24.39	3.57
2.0	64.96	12.38	5.25
3.0	49.19	9.48	5.18
4.0	17.42	5.98	2.91

Note: Extracts made from YNN27 (pα1-TPA) cells were subjected to Endo H treatment over a period of time and t-PA activity was measured as described in materials and methods.

E. ^{14}C-Labeling and Purification of t-PA

In order to monitor glycosylation of t-PA we did a labeling experiment in which the glycoproteins were labeled with ^{14}C-glucosamine. A 10-mℓ culture of YNN27 (pα1-TPA) was grown to an OD_{600} of 1.00; 50 μCi of ^{14}C-glucosamine (309 mCi/m mol) were added to the culture. The culture was allowed to grow for another 12 hr before being harvested. In order to see if there was any "O"-linked glycosylations of t-PA, the labeling experiment was repeated, except that the label was added 30 min after the addition of monensin (50 μℓ of 1 m*M* stock solution). Monensin, it should be mentioned, blocks "O"-linked glycosylations without affecting "N"-linked carbohydrate additions.

Labeled t-PA was purified as follows. Cell extracts, made by lysing cells with glass beads as described, were applied to a column in which antibodies raised against t-PA from Bowes melanoma cells are coupled to Sepharose® (American Diagnostica). The column was first washed with 0.1 *M* NaCl, 0.22 *M* PO_4 buffer, pH 7.3, followed by a wash with 10 mℓ 0.25 *M* potassium thiocyanate solution made in the wash buffer. t-PA was eluted with 1 *M* potassium thiocyanate. Potassium thiocyanate was removed by dialysis and t-PA was concentrated by burying the dialysis bag in polyethylene glycol. Half of the concentrated material was treated with Endo H and the treated and untreated samples analyzed on an SDS-PAGE gel followed by autoradiography.

F. t-PA Activity Gel

Instead of labeling t-PA, one can size the material synthesized in yeast by running an activity gel. t-PA was purified on an affinity column as described in Section E. The purified material was run on a Laemmeli gel in the absence of β-mercaptoethanol, but the samples contained 2% SDS. t-PA purified from Bowes melanoma cells and Chinese hamster ovary (CHO) cells was used as a marker. After electrophoresis SDS was exchanged with 2.5% Triton® X-100 for 15 min and the gel was rinsed in water for 20 min. The gel was then placed on a casein underlay and incubated at 37°C in a humidified atmosphere until clear zones developed. The casein underlay gel consisted of 1.3% agar, 2% casein, 10 m*M* Tris, pH 8.0, and 0.05 U/mℓ of plasminogen. All ingredients except plasminogen were mixed, agar was melted, and the mixture was cooled to 55°C. Plasminogen was then added and the gel was poured.

IV. RESULTS

A. Synthesis of t-PA Activity by Yeast Cells Transformed with the Plasmid pα1-TPA

Our first objective was to see whether the yeast cells would produce any t-PA activity after being transformed with the plasmid pα1-TPA described in Figure 1. One of the criteria that has to be satisfied was that such an activity has to show plasminogen dependence. Also, one has to take into account the possibility that a yeast protease in the cell extracts could either activate plasminogen to plasmin or cleave the chromogenic substrate, thus giving a false positive. Results of an experiment designed to distinguish between these possibilities are depicted in Figure 3. It is clear that there are no endogenous t-PA-like plasminogen activators in yeast. t-PA activity that was plasminogen dependent was detected only in cell extracts that were prepared from cells harboring the plasmid pα1-TPA. There was very little background activity in the extracts made from cells without the t-PA gene, and in extracts where no plasminogen was added.

B. t-PA Activity in the Extracellular Fluid

Since the MFα1 signal sequences would target the t-PA to the yeast secretory pathway, we monitored the extracellular fluid for any t-PA activity. Most of the time very little activity was detected in the medium. Occasionally, however, low levels of t-PA activity were detected, as shown in Figure 4. It is likely that the activity found in the medium was due to cell lysis. Alternatively, it could be the result of secretion, albeit, inefficient. We have not distinguished between these two possibilities yet.

C. Kinetics of t-PA Synthesis

Although the MFα1 promoter can be regulated in certain mutants,[9] it is constitutive in normal cells of the α-mating type. Thus, one would expect t-PA to be continuously synthesized in the logarithmic phase of the growth cycle. That such was the case is shown in Figure 5. t-PA activity levels off when the cells reach early stationary phase. When the cells

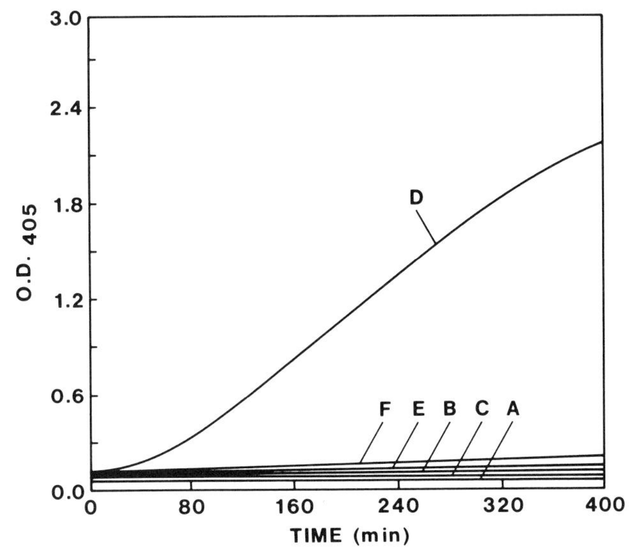

FIGURE 3. Expression of t-PA in YNN27 (pα1-TPA) cells. Cell extracts made from YNN27 (pα1-TPA) cultures and controls were monitored for the presence of t-PA as described in materials and methods. The various curves are identified as follows: (A) cell extract made from YNN27 cells carrying the plasmid pα1-ADHT (Figure 1B) minus plasminogen; (B) cell extracts as in (A) plus plasminogen; (C) cell extracts made from YNN27 (pα1-TPA) minus plasminogen; (D) cell extracts as in (C) plus plasminogen; (E) culture fluid from YNN27 (pα1-TPA) minus plasminogen; (F) culture fluid from YNN27 (pα1-TPA) plus plasminogen.

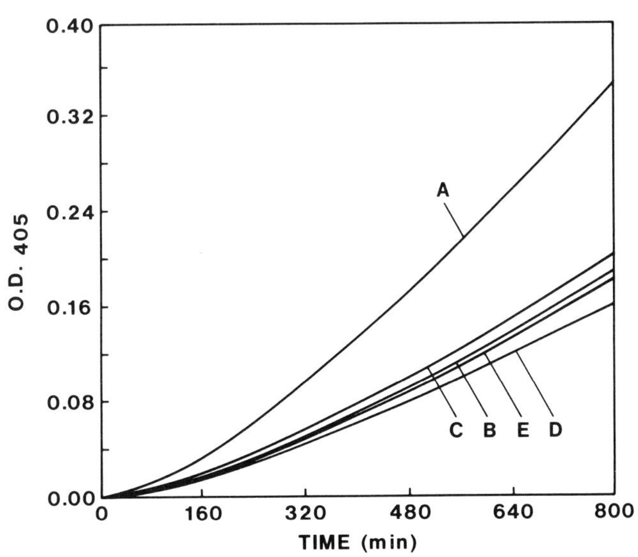

FIGURE 4. t-PA activity in the extracellular fluid. t-PA activity was not detected reproducibly in the culture medium. This figure shows the results of one of the experiments in which the kinetics of t-PA synthesis was being monitored. Low levels of activity in culture fluid were detected in this experiment; 100 μℓ of the fluid was monitored for the activity in each sample as described under materials and methods. The samples were taken at the following times: (A) 3 hr, (B) 26 hr, (C) 6 hr, (D) 9 hr, and (E) 12 hr.

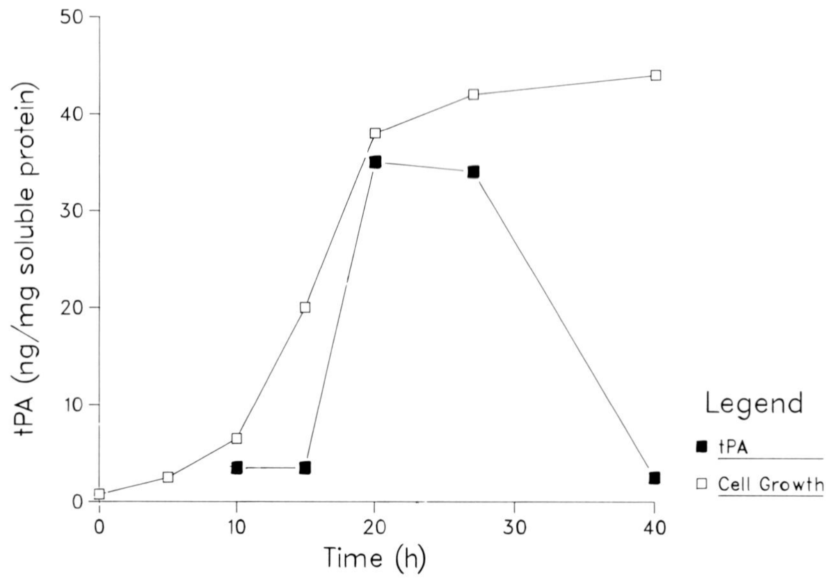

FIGURE 5. Kinetics of t-PA synthesis. The diagram depicts the synthesis of t-PA from the MFα1 promoter during the growth cycle. The synthesis of t-PA parallels the growth of cells, tapers off in early stationary phase, followed by a drop in the late stationary period.

enter the late stationary phase, t-PA activity starts to drop, probably due to proteolytic degradation.

We also tried to quantitate the amount of active t-PA synthesized in cells transformed with pα1-TPA. For this purpose a standard curve was developed using known amounts of purified t-PA. Details are described in materials and methods and Figure 2. Using such a standard curve we concluded that our current vector constructs were producing 15 to 30 ng/ mℓ/OD_{600} of active t-PA. However, this could be an underestimate (see the following section for more details).

D. Characterization of "Yeast" t-PA

Yeast, a eukaryote, is known to glycosylate a number of its proteins. It was logical to ask the question of whether it would also glycosylate t-PA, which has four N-linked glycosylation sites, although one of the four sites has *asn-pro-ser* which may not be glycosylated. Furthermore, we would have liked to know whether the MFα1 signal peptide was cleaved or not, as was the case during α-factor maturation.[6,7] To distinguish between various possibilities, first we ran an activity gel shown in Figure 6. "Yeast" t-PA was partially purified on an affinity column (see materials and methods) and ran on an activity gel along with t-PA purified from Bowes melanoma cells. It is clear from Figure 6 that "yeast" t-PA is of similar MW as the t-PA from mammalian cells. This result raised the possibilities that either the "yeast" t-PA is glycosylated and the signal peptide is cleaved or it is unglycosylated and has the signal peptide still attached.

That "yeast" t-PA is glycosylated is shown in Figure 7. ^{14}C-labeled t-PA was purified on an affinity column. Half the material was treated with Endo H as described. Endo H-treated and -untreated material was run on an SDS-gel followed by autoradiography. The result indicates that "yeast" t-PA is glycosylated and, most likely, the signal peptide is processed. Absolute confirmation of the latter possibility can only be done by NH_2-terminal sequencing of the purified "yeast" t-PA.

Now that it was established that "yeast" t-PA is glycosylated, we conducted another

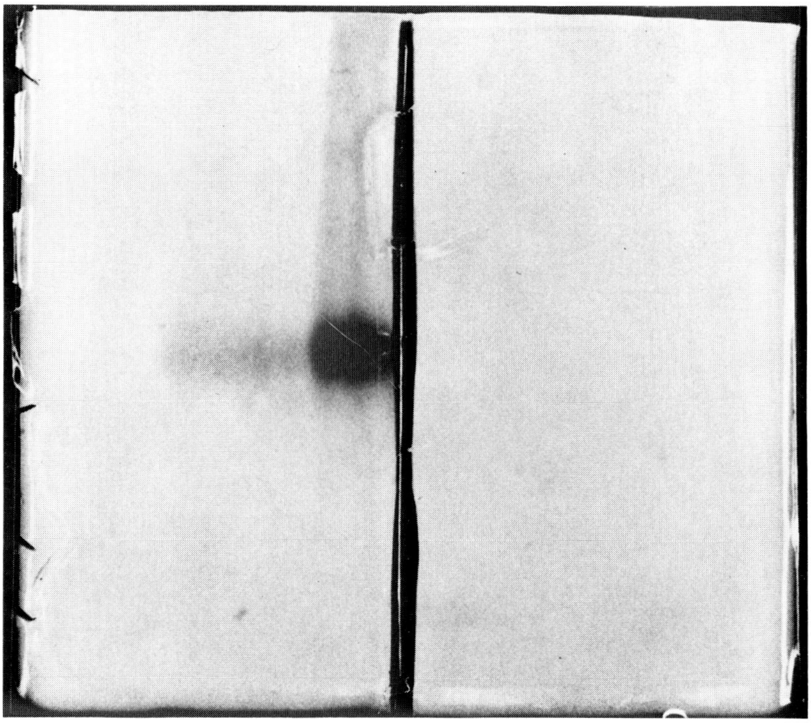

FIGURE 6. Sizing "yeast" t-PA on an activity gel. t-PA purified from mammalian cells and YNN27 (pα1-TPA) cells was run on an acrylamide gel followed by overlay onto an agarose gel as described under materials and methods. (a) "Yeast" t-PA; (b) t-PA from CHO cells, transformed with a vector carrying a cDNA copy of t-PA; and (c) t-PA from Bowes melanoma cells. The blank half is the control where no t-PA was present.

labeling experiment designed to probe for any "O"-linked glycosylations of t-PA. The drug monensin blocks "O"-linked glycosylations. When the t-PA labeled in the presence of monensin was analyzed on a gel, it was clear that the material was identical to the t-PA characterized previously (Figure 7), suggesting only "N"-linked glycosylation.

In some instances it has been shown that yeast heavily glycosylates some of its proteins, e.g., invertase.[14] Since "yeast" t-PA is glycosylated, we wanted to have some idea whether glycosylation or, perhaps, hyperglycosylation would affect the activity of the t-PA molecule. Cell extracts made from YNN27 (pα1-TPA) cells were treated with Endo H as described and monitored for t-PA activity. It is clear from Table 1 that the Endo H treatment results in a three- to fivefold increase in "yeast" t-PA activity. Also, there appears to be degradation of t-PA in cell lysates during Endo H treatment. However, the difference in activity level in treated and untreated extracts is maintained.

V. DISCUSSION

As mentioned in the beginning, t-PA is a complex protein containing a large number of cysteine residues.[8] Some of these residues are no doubt involved in the formation of specific disulfide bridges, perhaps stabilizing complex domains of the molecule. Thus, for t-PA to retain activity, and perhaps specificity, proper folding is of paramount importance. Experience with high level expression of some mammalian proteins in *E. coli* has shown that

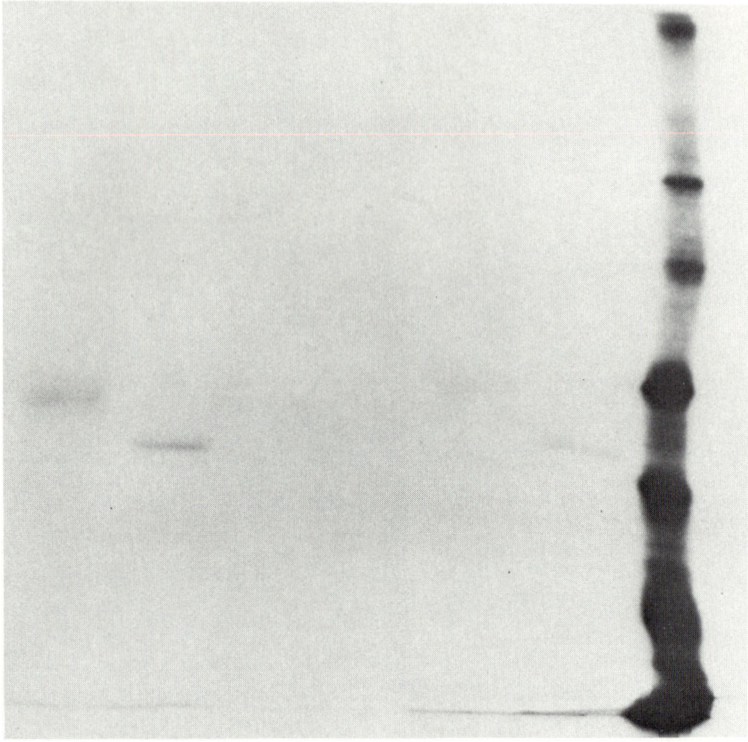

FIGURE 7. Glycosylation of "yeast" t-PA. This figure shows the results of experiments which proved conclusively that "yeast" t-PA was glycosylated. ^{14}C-labeled t-PA was purified, Endo H treated, and run on an acrylamide gel as described in materials and methods. Lanes are as follows: (a) ^{14}C-labeled t-PA without Endo H treatment; (b) ^{14}C-labeled t-PA treated with Endo H; (c) t-PA labeled with ^{14}C-glucosamine in the presence of monensin and not treated with Endo H; (d) material as in (c) but treated with Endo H.

the cytoplasmic environment is too reducing for these proteins to fold properly.[15,16] In eukaryotes it seems that the disulfide bonds form rapidly co- or posttranslationally in the lumen of the endoplasmic reticulum.[17,18] Recently, the presence of the enzyme protein disulfide isomerase, which is thought to promote the formation of intra- and intermolecular disulfide bonds in the endoplasmic reticulum, has been reported.[19,20] Yeast being a eukaryote probably has such an enzyme. These observations lead us to target the t-PA to the yeast secretory pathway using the MFα1 promoter and the "pre-pro" sequences. As the results indicate, this strategy led to the synthesis of active t-PA by yeast cells transformed with a plasmid carrying the t-PA gene. The activity was shown to be totally plasminogen dependent. Very little t-PA activity was, however, detected in the culture fluid.

We have characterized the "yeast" t-PA with respect to its molecular weight and post-translational modifications. The combined results of the experiments shown in Figures 6 and 7 clearly establish that t-PA is glycosylated and is of similar molecular weight as the t-PA isolated from Bowes melanoma cells. The result also implies that the MFα1 "pre-pro" sequences are cleaved from the t-PA precursor. Whether the cleavage is proper and complete can only be established by NH_2-terminal sequencing of "yeast" t-PA.

An interesting result of the Endo H treatment of "yeast" t-PA was the increase in the level of t-PA activity. At present we do not know if the increase in activity is due to activation of hyperglycosylated material that may otherwise be inactive or is due to the removal of carbohydrate moieties from the already active or partially active material. Further investigations are underway to distinguish between various possibilities.

Presently, the amount of active t-PA made in YNN27 (pα1-TPA) cells is rather low. It is likely that continuous synthesis of t-PA from a constitutive promoter like MFα1 is toxic to the cells. There is some evidence that the cells are under stress while making t-PA. For example, repeated subculturing of the t-PA-producing cells results in lower and lower yields of active t-PA. When these cells were plated on YMMCA, two colony types, small and large, appeared on the plates. The small colony types did not produce detectable levels of t-PA and did not grow in YMM without the addition of casamino acids. Use of regulated promoters like GAL1 and ADHII or of hosts in which MFα1 can be regulated should allow one to circumvent some of these problems.[9,21,22] Also, a yeast strain capable of secreting t-PA in the medium would, perhaps, ease some of the constraints.

While the levels of t-PA expression can be manipulated with the use of suitable hosts and vectors, a more serious problem for t-PA produced in yeast would be its antigenicity. Whereas the "N"-linked core glycosylations added by yeast are the same as in higher eukaryotes, the secondary extensions added in the golgi are different and mostly in the form of mannose residues. The high mannose content of "yeast" t-PA could perhaps be decreased by using certain *mnn* or *sec* mutants,[23,24] but whether it would minimize or eliminate the antigenicity of the material remains to be seen.

REFERENCES

1. **Rijkin, D. C. and Collen, D.,** Purification and characterization of the plasminogen activator secreted by human melanoma cells in culture, *J. Biol. Chem.,* 256, 7035, 1981.
2. **Matsuo, O., Rijkin, D. C., and Collen, D.,** Comparison of the relative fibrinogenolytic, fibrinolytic and thrombolytic properties of tissue plasminogen activator and urokinase in vitro, *Throm. Haemostasis,* 45, 225, 1981.
3. **Pennica, D., Holmes, W., Kohr, W., Harkins, R., Vehar, G., Ward, C., Bennett, W., Yelverton, E., Seeburg, P., Hayneker, H., Goeddel, D., and Collen, D.,** Cloning and expression of human tissue-type plasminogen activator cDNA in *E. coli, Nature,* 301, 214, 1983.
4. **Marotti, K. R.,** unpublished, 1983.
5. **Tomich, C.-S. C. and Marotti, K. R.,** unpublished, 1984.
6. **Kurjan, J. and Herskowitz, I.,** Structure of a yeast pheromone gene (MFα): a putative α-factor precursor contains four tandem repeats of mature α-factor, *Cell,* 30, 933, 1982.
7. **Singh, A., Chen, E., Lugovoy, J., Chang, C., Hitzeman, R., and Seeburg, P.,** *Saccharomyces cerevisiae* contains two discrete genes coding for the α-factor pheromone, *Nucleic Acids Res.,* 11, 4049, 1983.
8. **Bitter, G., Chen, K., Banks, A., and Lai, P.,** Secretion of foreign proteins from *Saccharomyces cerevisiae* directed by α-factor, *Proc. Natl. Acad. Sci. U.S.A.,* 81, 5330, 1984.
9. **Brake, A., Merryweather, J., Coit, D., Heberlein, U., Masiarz, F., Mullenbach, G., Urdea, M., Valenzuela, P., and Barr, P.,** α-Factor directed synthesis and secretion of mature foreign proteins in *Saccharomyces cerevisiae, Proc. Natl. Acad. Sci. U.S.A.,* 81, 4642, 1984.
10. **Sherman, F. and Fink, G.,** *Methods in Yeast Genetics,* Cold Spring Harbor Laboratory, Cold Spring Harbor, N.Y., 1979, 61.
11. **Miller, J.,** *Experiments in Molecular Genetics,* Cold Spring Harbor Laboratory, Cold Spring Harbor, N.Y., 1972, 431
12. **Beggs, J. D.,** Transformation of yeast by a replicating hybrid plasmid, *Nature,* 275, 104, 1978.
13. **Tang, J., Li, S., McGray, P., and Vechhio, A.,** Current status of activity assay for tissue plasminogen activator, in *Enzyme Engineering,* Vol. 7, Laskin, A. I., Tsao, G. T., and Wingard, L. B., Jr., Eds., New York Academy of Sciences, New York, 1984, 536.
14. **Taussig, R. and Carlson, M.,** Nucleotide sequence of the yeast SUC2 gene for invertase, *Nucleic Acids Res.,* 11, 1943, 1983.
15. **Wetzel, R., Kleid, D. G., Crea, R., Heyneker, H., Yansura, D. G., Hirose, T., Kraszenski, A., Riggs, A., Itakura, K., and Goeddel, D.,** Expression in *Escherichia coli* of a chemically synthesized gene for a "mini-C" analog of human proinsulin, *Gene,* 16, 63, 1981.

16. **Cabilly, S., Riggs, A., Pande, H., Shively, J., Holmes, W., Rey, M., Perry, L., Wetzel, R., and Heyneker, H.,** Generation of antibody activity from immunoglobulin polypeptide chains produced in *Escherichia coli, Proc. Natl. Acad. Sci. U.S.A.,* 81, 3273, 1984.
17. **Bergman, L. W. and Kuehl, W. M.,** Formation of intermolecular disulphide bonds on nascent immunoglobulin polypeptides, *J. Biol. Chem.,* 254, 5690, 1979.
18. **Peters, T. and Davidson, L. K.,** The biosynthesis of rat serum albumin. In vivo studies on the formation of the disulphide bonds, *J. Biol. Chem.,* 257, 8847, 1982.
19. **Edman, J., Ellis, L., Blacher, R., Roth, R., and Rutter, W.,** Sequences of protein disulphide isomerase and implications of its relationship to thioredoxin, *Nature,* 317, 267, 1985.
20. **Goldberger, R. F., Epstein, C. J., and Anfinsen, C. B.,** Acceleration of reactivatin of reduced bovine pancreatic ribonuclease by a microsomal system from rat liver, *J. Biol. Chem.,* 238, 628, 1983.
21. **Johnston, M. and Davis, R. W.,** Sequences that regulate the divergent GAL1-GAL10 promoter in *Saccharomyces cerevisiae, Mol. Cell. Biol.,* 4, 1440, 1984.
22. **Russell, D., Smith, M., Williamson, V., and Young, E. T.,** Nucleotide sequence of the yeast Alcohol Dehydrogenase II gene, *J. Biol. Chem.,* 258, 2674, 1983.
23. **Ballous, C. E.,** Yeast cell wall and cell surface, in *The Molecular Biology of the Yeast Saccharomyces,* Strathern, J. N., Jones, E. W., and Broach, J. R., Eds., Cold Spring Harbor Laboratory, Cold Spring Harbor, N.Y., 1982, 335.
24. **Schekman, R. and Novick, P.,** The secretory process and yeast cell-surface assembly, in *The Molecular Biology of the Yeast Saccharomyces,* Strathern, J. N., Jones, E. W., and Broach, J. R., Eds., Cold Spring Harbor Laboratory, Cold Spring Harbor, N.Y., 1982, 361.

Chapter 15

BIOSYNTHESIS OF ISOLEUCINE AND VALINE IN *SACCHAROMYCES* GENE STRUCTURE AND REGULATION

Steen Holmberg, Jens G. L. Petersen, Torsten Nilsson-Tillgren, Claes Gjermansen, and Morten C. Kielland-Brandt

TABLE OF CONTENTS

I. INTRODUCTION

In the yeast *Saccharomyces* five enzymes catalyze the biosynthesis of isoleucine and valine.[1] The first enzyme in the pathway for isoleucine is threonine deaminase, an allosteric enzyme converting threonine to α-ketobutyrate. The other four enzymes are shared by the biosynthetic pathway for valine, converting α-ketobutyrate to isoleucine and pyruvate to valine. The immediate precursor for valine, α-ketoisovaleric acid, serves as the starting point for the synthesis of leucine.[2] Threonine deaminase as well as the three subsequent enzymes are found associated with the mitochondria.[3]

Threonine deaminase is coded for by *ILV1*, located on chromosome V.[4] Mutations in *ILV1* lead to auxotrophy for isoleucine. The acetohydroxyacid synthetase is coded for by *ILV2*, acetohydroxyacid reductoisomerase by *ILV5*, and dihydroxyacid dehydrase by *ILV3*. Mutations in any of the latter three genes lead to auxotrophy for both isoleucine and valine. *ILV2* is located on chromosome XIII, *ILV3* is on chromosome X, and *ILV5* on chromosome XII.[4,5] No gene has been assigned for the last enzymatic step, which is a transamination.

The amounts of the enzymes in the pathways are regulated. Compared to the levels found in cells grown in minimal medium, repression is observed by the addition of isoleucine, valine, and leucine (resembling multivalent repression in *Escherichia coli*).[6-8] This multivalent repression has also been reported to be under the influence of the *ILV1* gene product by an unknown mechanism.[9]

We have undertaken an investigation of the structure of the isoleucine-valine genes and their regulation. Recently, the four genes have been cloned from *S. cerevisiae* and *ILV1*, *ILV2*, and *ILV5* sequenced.[10-14] Furthermore, we have employed internuclear chromosome transfer mediated by the *kar1* mutation, combined with Southern hybridization and molecular cloning, in our analysis of the *ILV* genes of a *S. carlsbergensis* yeast used for the manufacture of lager beer.[15-17]

II. REGULATION OF *ILV1* AND *ILV5*

At least 28 different genes coding for enzymes involved in different amino acid biosynthetic pathways are under general control of amino acid biosynthesis. To analyze whether the synthesis of threonine deaminase, coded for by *ILV1*, is also subject to this control, *S. cerevisiae* wild-type cells (strain X2180-1A) were starved for tryptophan by growth in minimal medium supplemented with the tryptophan analog 5-methyl tryptophan (5-MT). The specific activity of the enzyme, as well as the amount of the mRNA, was compared to the levels found after growth in minimal medium. Table 1 shows that when 5-MT is added, threonine deaminase enzyme activity is derepressed about twofold. The threonine deaminase mRNA increases by the same factor. These results indicate that *ILV1* is subject to general control. This is supported by our results with mutants constitutively derepressed for the general control system (*gcd* mutants, Table 1). The enzyme activity and mRNA level are elevated in both *gcd2* (strain RH558-1)[18] and *gcd3* (strain RH772)[18] cells to the same extent as in wild-type cells that are starved for amino acids. Furthermore, in the 5′ flanking region of *ILV1*, three copies of the general control consensus sequence TGACT are present.[12,19] It is presumed that the *GCN4* product interacts with these sites to activate transcription under conditions of amino acid starvation.[20]

Threonine deaminase, as well as the other enzymes of the pathways, has been reported to be repressed by isoleucine, valine, and leucine (multivalent repression).[6,7] We analyzed threonine deaminase with respect to this regulation. Table 1 shows that the steady-state level of threonine deaminase in wild-type cells is repressed to 80% when isoleucine, valine, and leucine are added to minimal medium, in agreement with earlier results.[6] The repression appears to be transcriptional. Whether multivalent repression is related to general control of amino acid biosynthesis is discussed below.

Table 1
STEADY-STATE LEVELS OF *ILV1* ENZYME AND *ILV1* mRNA UNDER VARIOUS GROWTH CONDITIONS[30]

Strain		Medium	Enzyme level[a]	mRNA level[b]
X2180-1A	wild-type	MA[c]	1	1
		MA + 5-MT	1.8	1.9
		MA + Ile + Val + Leu	0.8	0.7
RH558-1	*gcd2*	MA	1.7	2.0
RH772	*gcd3*	MA	1.7	2.0

[a] Threonine deaminase activity was measured according to de Robichon-Szulmajster and Magee[31] in Triton® X-100 permeabilized cells.[32] The ratios are the means of two to four independent experiments.
[b] The relative *ILV1* transcript levels were compared by Northern analysis using total RNA. The filters were hybridized with ^{32}P-labeled probes containing the *ILV1* and the *URA3* gene. The latter transcript was used as an internal control for equal amounts of total RNA. The ratios are the means of two to four independent experiments.
[c] Minimal medium with ammonia as nitrogen source.

Table 2
STEADY-STATE LEVELS OF *ILV5* ENZYME AND *ILV5* mRNA UNDER VARIOUS GROWTH CONDITIONS[13]

Strain		Medium	Enzyme level[a]	mRNA level[b]
X2180-1A	wild-type	MA	1	1
		MA + 5-MT	0.9	1.1
		MA + Ile + Val + Leu	0.6	0.6
		MP[c]	1.1	ND[d]
		MP + 5-MT	1.4	ND
RH558-1A	*gcd2*	MA	1.4	ND
RH772	*gcd3*	MA	1.4	ND

[a] As in Table 1, but reductoisomerase activity was determined according to Magee and Hereford.[7]
[b] As in Table 1, but using *ILV5* and *URA3* DNA as probes.
[c] Minimal medium with proline as nitrogen source.
[d] ND = not determined.

We have also analyzed the regulation of the acetohydroxyacid reductoisomerase which is coded for by the *ILV5* gene. Again, in order to ask whether *ILV5* is regulated by the general control system, enzyme activity and the mRNA level were compared in wild-type cells grown in minimal medium with or without the addition of 5-MT (Table 2). In this case starvation for tryptophan in minimal medium does not derepress *ILV5* when ammonia is the nitrogen source. However, the amino acid limitation induced by using proline as the nitrogen source caused a slight derepression of the enzyme activity. Increased amino acid imbalance brought about by adding 5-MT to this medium resulted in a further derepression. In addition, derepression to the same level is seen in the *gcd2* and *gcd3* mutants grown in minimal medium with ammonia. From this we conclude that *ILV5* is subject to general amino acid control. In agreement with this, we have found several TGACT elements in the 5′ noncoding region of *ILV5*.[13] As in the case with threonine deaminase, the multivalent repression by isoleucine, valine, and leucine of the reductoisomerase activity appears to be transcriptionally regulated (Table 2).

Table 3
STEADY-STATE LEVELS OF *HIS4C* ENZYME AND *HIS4* mRNA IN STRAIN X2180-1A UNDER VARIOUS GROWTH CONDITIONS. FOR COMPARISON THE DATA FOR *ILV1* IS INCLUDED[30]

	HIS4		*ILV1*	
Medium	*HIS4C* enzyme level[a]	mRNA level[b]	Enzyme level[c]	mRNA level[c]
MA	1	1	1	1
MA + 5-MT	1.7	1.7	1.8	1.9
MA + Ile + Val + Leu	0.8	0.7	0.8	0.7

[a] Histidinol dehydrogenase activity was measured as described by Fink and Styles.[33]
[b] As in Table 1, but using *HIS4* and *URA3* DNA as probes.
[c] Data from Table 1.

In conclusion, the results presented above show that both *ILV1* and *ILV5* are subject to general control. Also, both gene products increase in media lacking one or more of the branched-chain amino acids.

III. EVIDENCE THAT MULTIVALENT REPRESSION AND GENERAL CONTROL ARE RELATED

If multivalent repression is specific for the biosynthesis of isoleucine and valine, it should not be observable in the expression of a gene in another amino acid biosynthetic pathway controlled by the general amino acid system. The *HIS4* gene codes for three enzymatic steps leading to the synthesis of histidine[21] and is composed of the *HIS4 A, B, C* regions. Specific regulation exerted by histidine alone has not been found.[22] The data for *HIS4* are summarized and compared to *ILV1* in Table 3. Histidinol dehydrogenase activity encoded by the *HIS4C* region is derepressed when cells of the wild-type strain X2180-1A are starved for tryptophan by growth in minimal medium containing 5-MT. The *HIS4* mRNA is derepressed to the same extent and the amounts of enzyme and mRNA correspond well to what is expected under derepressed conditions. When isoleucine, valine, and leucine are added to minimal medium the levels of both *HIS4C* enzyme and *HIS4* mRNA are repressed, and the degree of repression is comparable to what is seen for *ILV1*, showing that multivalent repression is not pathway specific and probably mediated by general control. These results suggest that wild-type cells growing in minimal medium are limited for one or more of the branched-chain amino acids leading to noncoordinate partial derepression of the enzymes that are under general control. Indeed, the growth rate of wild-type cells is decreased when all three amino acids are absent from the medium.[23] Our suggestion is consistent with the following findings. First, derepression of *ILV1* and *ILV5* as well as *HIS4* is seen when wild-type cells grown in minimal medium supplemented with isoleucine, valine, and leucine are starved for tryptophan by feeding 5-MT. Second, little, if any, multivalent repression of *ILV1* and *ILV5* takes place when proline is the nitrogen source.

In summary, multivalent repression is not pathway specific, but is at least to some extent mediated by the general amino acid control system.

```
Gene    Deduced amino-terminal amino acid sequence

                 10        20        30        40        50        60
ILV1    MSATLLKQPLCTVVRQGKQSKVSGLNLLRLKAHLHRQHLSPSLIKLHSELKLDELQTDNT
              -       -  -  -       - - - -- -      - -   - **   *

                 10        20        30        40        50        60
ILV2    MIRQSTLKNFAIKRCFQHIAYRNTPAMRSVALAQRFYSSSSRYYSASPLPASKRPEPAPS
          -    -    --   -   -      -     -      -          -- *

                 10        20        30        40        50        60
ILV5    MLRTQAARLICNSRVITAKRTFALATRAAAYSRPAARFVKPMITTRGLKQINFGGTVETV
          -    -     -    --      -     -   -  -     -  -        *
```

FIGURE 1. Comparison of the amino terminal region of the *ILV1*,[12] *ILV2*,[14] and *ILV5*[13] polypeptides. The first 60 amino acids deduced from the DNA sequences are shown. The basic amino acids — arginine = R, lysine = K, and histidine = H — are underlined. The acidic amino acids — glutamic acid = E and aspartic acid = D — are marked by ★.

IV. THE *ILV* GENES OF *S. CEREVISIAE*

As already mentioned, the *ILV1, ILV2, ILV3,* and *ILV5* genes have been cloned by functional complementation of the corresponding yeast mutations.[10,11] The identity of the cloned genes was established by their capability to complement a collection of *ilv* mutant alleles and by site-directed integration at the *ilv* loci. *ILV2* has also been cloned by Falco et al.[24] by its ability to confer low-level resistance to the herbicide sulfometuron methyl on a high copy number plasmid. We have studied *ILV1* and *ILV5* in detail. The DNA sequence of the *ILV1* region revealed a 1728-base pair (bp)-long open translational reading frame.[12] The molecular weight of the primary gene product is larger than the published value for the enzyme monomer, suggesting that a peptide of 15,000 daltons is cleaved off in an intracellular processing of the enzyme. Characterization of the *ILV1* mRNA indicated that the full 1728-bp open reading frame is translated.[25] DNA sequencing of the *ILV5* region has revealed a translational open reading frame of 1185 bp, giving a primary gene product with a molecular weight of 44,280. Northern analysis has identified a mRNA of about 1500 nucleotides, which covers the reading frame.[13]

All the enzymes of the pathways, with the possible exception of the transaminase, are found sequestered to the mitochondria.[3] In Figure 1 is shown the deduced amino acid sequence of the amino terminal part of the *ILV1, ILV2,* and *ILV5* gene products. The data for *ILV2* are from Falco et al.[14] The amino terminal region has been implicated in the targeting of proteins to the mitochondria.[26] All three polypeptides have a highly basic amino terminus. In the *ILV1* gene product there are no acidic residues among the first 47 amino acids, but there are 3 arginines, 5 lysines, and 4 histidines. Likewise, in the *ILV5* gene product the first acidic residue is at position 58, and among the first 49 amino acids 8 arginines and 3 lysines are found. Also, the amino terminus of the *ILV2* gene product is highly basic, having 7 arginines, 3 lysines, and 1 histidine among the first 54 residues. Thus, it is likely, as is the case with other mitochondrially imported proteins, that the amino end of all three polypeptides has a role in their mitochondrial import. There is no part with extended amino acid sequence homology between this region of the three polypeptides. It remains to be determined whether amino terminal residues are cleaved off in the process of transport.

V. HOMEOLOGOUS *ILV* GENES IN *S. CARLSBERGENSIS*

We are also studying the genetic and molecular organization of the *ILV* genes in a lager yeast, *S. carlsbergensis* strain 244. As with many industrial yeasts, strain 244 shows poor

Table 4
CROSSING-OVER FREQUENCES (cM) OBTAINED BY TETRAD ANALYSIS OF STRAINS DISOMIC OR TRISOMIC FOR CHROMOSOME V[17]

	can1		*ura3*		*cen5*		*his1*		*ilv1*		*rad3*
cerevisiae[a]		42.2		7.8		43.9		14.9		84.1	
Homolog *carlsbergensis* addition line		ND[b]		10.2		41.3		16.9		42.5	
Homolog *carlsbergensis* substitution line		31.1		7.8		37.0		22.4		87.1	
Homeolog *carlsbergensis* substitution line		1		0		0		0		0	

[a] Genetic distances according to Mortimer and Schild.[4]
[b] ND = not determined.

sporulation and poor spore viability. Thus, for the genetic analysis we have employed internuclear chromosome transfer mediated by the *kar1* mutation,[27] which greatly reduces the frequency of karyogamy after zygote formation. One or a few chromosomes can be transferred from one nucleus to the other in heterokaryons formed in *kar1* × *KAR1* crosses.[28,29] In this way strains have been constructed that carry one chromosome from *S. carlsbergensis* in a genetically well-marked *S. cerevisiae* strain. Occasionally, the *S. cerevisiae* chromosome is lost, resulting in chromosome substitution lines in which the *S. cerevisiae* chromosome is replaced by a *S. carlsbergensis* chromosome. With this method we have isolated *S. carlsbergensis* chromosomes which are functionally homologous to chromosomes III,[15] IV, V,[17] VII, XII, and XIII of *S. cerevisiae*. As a typical example, *S. carlsbergensis* chromosome V, carrying *ILV1,* will be discussed in some detail. Chromosome V was transferred from meiotic progeny of strain 244 to a *kar1 S. cerevisiae* recipient with five markers on chromosome V covering the entire known map, from *can1* to *rad3*. The genetic analysis of chromosome V addition strains and substitution strains is summarized in Table 4. First, two different types of chromosome V were transferred: one showed normal recombination with the *S. cerevisiae* chromosome V in all the intervals between *can1, ura3, cen5, his1, ilv1,* and *rad3,* whereas the other recombined only very rarely. Second, the construction of both types of substitution lines shows that either one is functionally homologous to chromosome V of *S. cerevisiae,* i.e., in addition to *URA3, HIS1, ILV1,* and *RAD3,* they contain all essential functions normally encoded by chromosome V. Third, in the analysis of the non-recombining substitution line, spore viability was between 80 and 90%, showing that proper disjunction at meiosis I takes place, although recombination between *can1* and *rad3* is less than 1%. Fourth, the absence of recombination is correlated with a different nucleotide sequence of the *S. carlsbergensis* genes. We have analyzed four different loci, *ilv1, ura3, cyc7,* and *can1,* by Southern hybridization using cloned *S. cerevisiae* DNA as probes. The analysis of *ILV1* is shown in Figure 2A using a 6-kb HindIII-SalI fragment as probe.[10] *S. cerevisiae* DNA digested with HindIII (lane a) gives two strongly hybridizing bands of 9.4 and 3.1 kb, respectively, whereas weak hybridization to a band of 2.6 kb is seen with DNA from the nonrecombining substitution strain (lane d). The pattern obtained with a disomic strain having the recombining chromosome (lane c) is identical to that of *S. cerevisiae* DNA (lane a). Furthermore, strain 244 shows a combination of both patterns (lane b). Similar results were obtained when *URA3, CYC7,* and *CAN1* DNA were used as probes (data not shown). These results confirm the genetic data showing the presence in *S. carlsbergensis* of two genetically different chromosomes V.

Recently, chromosomes XII and XIII, carrying *ILV5* and *ILV2,* have been analyzed. In

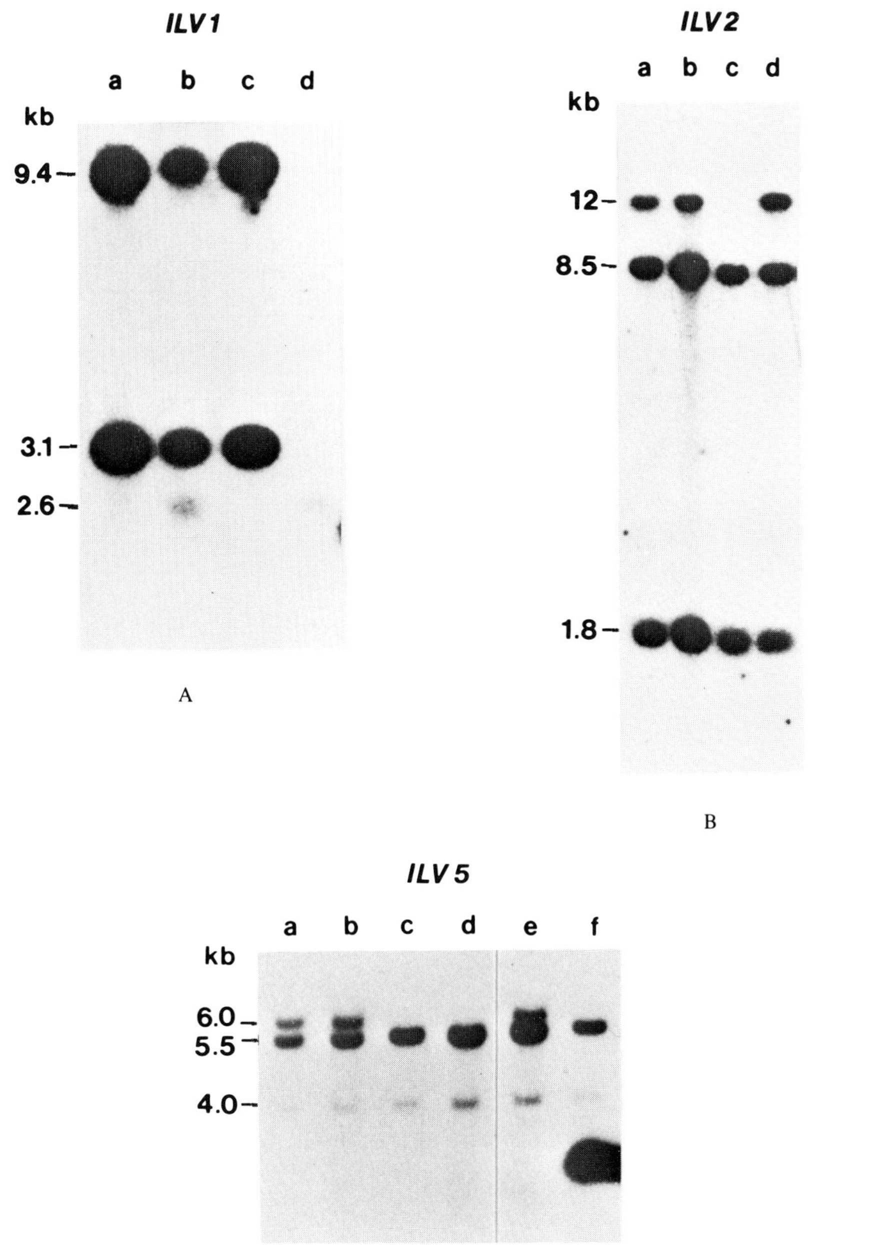

FIGURE 2. (A) Southern blot analysis of the *ILV1* region in *S. cerevisiae* strain C80-II-17 (lane a), *S. carlsbergensis* strain 244 (lane b), homologous chromosome V addition strain C80-1607 (lane c), and homeologous chromosome V substitution strain C80-1609 (lane d). The DNA was digested with HindIII and probed with a ^{32}P-labeled 6-kb HindIII-SalI fragment containing the *S. cerevisiae ILV1* gene.[10] (B) Southern blot analysis of the *ILV2* region in *S. carlsbergensis* strain 244 (lane a), meiotic offspring of strain 244, C80-CG110 (lane b), *S. cerevisiae strain* C82-1827 (lane c), and homeologous chromosome XIII addition strain C82-1950 (lane d). The DNA was digested with PvuII and probed with a 2.9-kb KpnI-HindIII fragment containing the *S. cerevisiae ILV2* coding region.[14] (C) Southern blot analysis of the *ILV5* region in *S. carlsbergensis* strain 244 (lane a), meiotic offspring of strain 244, C80-CG7 (lane b), *S. cerevisiae* strain S288C (lane C), *S. cerevisiae* recipient strain C82-1924 (lane d), homeologous chromosome XII addition strain C83-2300 (lane e), and homologous chromosome addition strain C83-2304 (lane f). The DNA was digested with ClaI and probed with a 2.3-kb BamHI-EcoRI fragment containing 90% of the *S. cerevisiae ILV5* coding region.[13]

```
                    1
S. carlsbergensis         T                  GA   G   C     A   G       TC          A    G  AGT
S. cerevisiae       ATG TCA GCT ACT CTA CTA AAG CAA CCA TTA TGT ACG GTT GTT CGG CAA GGT AAA CAG
S. cerevisiae       Met Ser Ala Thr Leu Leu Lys Gln Pro Leu Cys Thr Val Val Arg Gln Gly Lys Gln
S. carlsbergensis                           Arg             Ser Ala     Ser         Ser Arg Ser
                  58
                     G   G    C         C T              A                A   A   T       C   T
                    TCC AAA GTG TCT GGA TTG AAC CTT TTG AGA CTA AAG GCT CAT TTG CAC AGA CAA CAC
                    Ser Lys Val Ser Gly Leu Asn Leu Leu Arg Leu Lys Ala His Leu His Arg Gln His
                        Arg                             Lys             Gln             His
                  115
                    T A           T   A                   C   G         C                     C
                    CTG TCA CCT TCC TTG ATA AAA CTA CAC TCT GAA TTG AAA TTG GAT GAG CTG CAA ACT
                    Leu Ser Pro Ser Leu Ile Lys Leu His Ser Glu Leu Lys Leu Asp Glu Leu Gln Thr

                  172
                              T   G   C   T       C   G   A   G   A   G   C   G       C
                    GAT AAC ACC CCT GAT TAC GTC CGT TTA GTT TTA AGG TCC TCT GTA TAC GAT GTT ATT
                    Asp Asn Thr Pro Asp Tyr Val Arg Leu Val Leu Arg Ser Ser Val Tyr Asp Val Ile

                  229
                              A                           C           T     T G
                    AAT GAA TCT CCA ATC TCT CAA GGT GTA GGT TTG TCT TCC CGT CTA AAC ACG AAT GTC
                    Asn Glu Ser Pro Ile Ser Gln Gly Val Gly Leu Ser Ser Arg Leu Asn Thr Asn Val

                  286
                      T       G               T           G                               T
                    ATC TTG AAA AGA GAA GAT CTA TTG CCT GTT TTC TCT TTC AAG CTT CGT GGT GCC TAT
                    Ile Leu Lys Arg Glu Asp Leu Leu Pro Val Phe Ser Phe Lys Leu Arg Gly Ala Tyr

                  343
                                        C A
                    AAC ATG ATT GCC AAG TTG GAC GAT TCT CAA AGA AAC CAG GGT GTT ATT GCC TGT TCA
                    Asn Met Ile Ala Lys Leu Asp Asp Ser Gln Arg Asn Gln Gly Val Ile Ala Cys Ser

                  400
                      A   A   C   C                   T       A   C       T       G   C   G
                    GCT GGG AAT CAT GCC CAA GGT GTG GCC TTT GCT GCT AAA CAC TTG AAA ATA CCT GCT
                    Ala Gly Asn His Ala Gln Gly Val Ala Phe Ala Ala Lys His Leu Lys Ile Pro Ala

                  457
                      C       C           C                   C               C       C       G
                    ACT ATC GTT ATG CCT GTT TGT ACA CCA TCT ATT AAG TAT CAA AAT GTC TCG AGA TTA
                    Thr Ile Val Met Pro Val Cys Thr Pro Ser Ile Lys Tyr Gln Asn Val Ser Arg Leu

                  514
                                  T   T   T  T        T               A   C   A   A
                    GGG TCT CAA GTC GTC CTA TAT GGT AAC GAT TTT GAC GAG GCT AAG GCT GAA TGT GCC
                    Gly Ser Gln Val Val Leu Tyr Gly Asn Asp Phe Asp Glu Ala Lys Ala Glu Cys Ala
                                            Phe
                  571
                              A       A G     C A   A                                   C
                    AAA TTG GCT GAA GAG CGT GGC TTG ACG AAC ATT CCT CCT TTC GAT CAT CCT TAT GTC
                    Lys Leu Ala Glu Glu Arg Gly Leu Thr Asn Ile Pro Pro Phe Asp His Pro Tyr Val
                                        Gly
                  628
                          A                       C                                      T
                    ATT GCC GGT CAA GGT ACT GTA GCT ATG GAA ATC CTA AGA CAA GTA CGT ACC GCT AAT
                    Ile Ala Gly Gln Gly Thr Val Ala Met Glu Ile Leu Arg Gln Val Arg Thr Ala Asn
                                                                                        Val
```

FIGURE 3. Comparison of *S. cerevisiae* and *S. carlsbergensis ILV1* coding sequences. The first 684 nucleotides and 228 deduced amino acids are shown. For *S. carlsbergensis*, only nucleotide and amino acid differences are depicted.[34]

both cases the addition lines fell into two classes with respect to recombination in *S. cerevisiae*. A molecular analysis of *ILV2* is shown in Figure 2B. Again, a weak hybridization pattern specific for the nonrecombining chromosome XIII is obtained (lane c) with a *S. cerevisiae ILV2* probe, whereas the recombining chromosome has a pattern identical to that of *S. cerevisiae* (lanes a and b). Similar hybridization data are obtained for *ILV5* (Figure 3C).

In order to get a precise quantitation of the observed nucleotide difference and to compare both the coding region and the flanking regions with the corresponding DNA of *S. cerevisiae*,

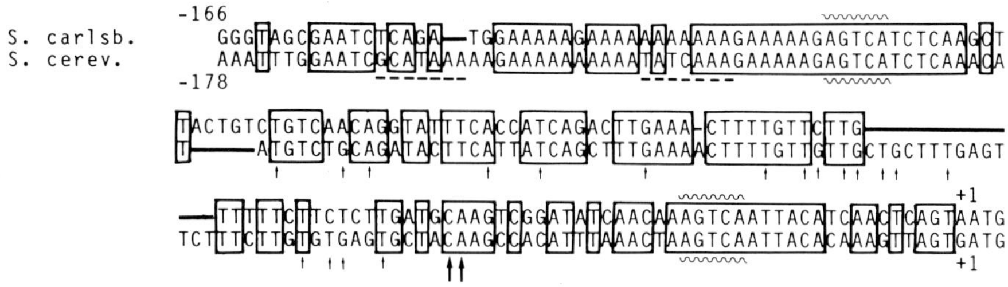

FIGURE 4. Comparison of DNA sequences of the upstream region of *S. cerevisiae*[12] and *S. carlsbergensis ILV1*.[34] Base pair gaps are indicated by unbroken lines. In the *S. cerevisiae* sequence mRNA initiation sites[25] are indicated by arrows. The general control sequence complements AGTCA are marked with an undulating line. Also indicated are TATA-like sequences (broken lines).[12]

both *ILV1* alleles of strain 244 have been cloned.[17] The DNA sequence of the 166-bp upstream region and 1266 bp of the coding region of the nonrecombining *ILV1* allele has been determined. In Figure 3 the first 684 nucleotides of the coding region of this *S. carlsbergensis* allele are compared to the corresponding sequence of *S. cerevisiae ILV1*. Also shown is the deduced amino acid sequence of the threonine deaminase. Within the first 141 bp the nucleotide inhomology is 20%. This results in 23 codon changes giving 11 substitutions among the 47 amino terminal amino acids. This is in contrast to the rest of the sequenced coding region where a nucleotide inhomology of 11% gives 118 codon changes, resulting in only 9 amino acid residue substitutions. It is interesting that the changes within the first 47 amino acids preserve the highly basic character of this region: three lysines are substituted with arginines, one arginine with a lysine, and a histidine to glutamine substitution is counterbalanced by changing a glutamine to histidine. This result may mean that the basic nature of the amino terminal region is functionally important, while other requirements to the amino acid sequence are relaxed, at least in part of the region.

The 5′ noncoding regions of the two *ILV1* alleles are compared in Figure 4. The 178 nucleotides 5′ to the ATG of the *S. cerevisiae ILV1* gene have been aligned with 166 nucleotides of the corresponding *S. carlsbergensis* DNA. Among the 70% conserved nucleotides in this region, there are sequence elements of possible significance for promoter function. The 5′ end of the *S. cerevisiae ILV1* mRNA displays a high degree of heterogeneity revealing 20 different transcription initiation sites spanning a region of 78 nucleotides.[25] None of the 20 different mRNAs is preferentially expressed in connection with the regulation of the gene. The two most prevalent transcripts comprise about half of the *ILV1* mRNA and initiate at −39 and −40 at the sequence CAAG[25] (Figure 4). Upstream of the initiation sites two TATA-like sequences can be recognized in the presented region. Also, three general control elements are found upstream of *ILV1*, two of which are present in the sequence depicted in Figure 4. In the *S. carlsbergensis ILV1* upstream DNA, the sequence CAAG is found at the same position relative to the ATG as in *S. cerevisiae*. Both general control sequences are conserved as are their relative positions, even though one of the elements is located within the transcribed region. None of the two TATA-like sequences of *S. cerevisiae* is conserved in the *S. carlsbergensis* DNA. It will be of interest to compare the transcription and regulation of these *ILV1* alleles.

ACKNOWLEDGMENTS

We thank D. von Wettstein for support and encouragement and critical reading of the manuscript. This work was supported by grants 16-3557.H-699 from the Danish National Research Council for Technical Sciences, and 133/001-84154 from Teknologirådet of the Danish Ministry of Industry to D. von Wettstein.

REFERENCES

1. **Kakar, S. N. and Wagner, R. P.,** Genetic and biochemical analysis of isoleucine-valine mutants of yeast, *Genetics,* 49, 213, 1964.
2. **Satyanarayana, T., Umbarger, H. E., and Lindegren, G.,** Biosynthesis of branched-chain amino acids in yeast: correlation of biochemical blocks and genetic lesions in leucine auxotrophs, *J. Bacteriol.,* 96, 2012, 1968.
3. **Ryan, E. D. and Kohlhaw, G. B.,** Subcellular localization of isoleucine-valine biosynthetic enzymes in yeast, *J. Bacteriol.,* 120, 631, 1974.
4. **Mortimer, R. K. and Schild, D.,** Genetic map of *Saccharomyces cerevisiae, Microbiol. Rev.,* 44, 519, 1980.
5. **Petersen, J. G. L., Kielland-Brandt, M. C., Holmberg, S., and Nilsson-Tillgren, T.,** Mutational analysis of isoleucine-valine biosynthesis in *Saccharomyces cerevisiae.* Mapping of *ilv2* and *ilv5, Carlsberg Res. Commun.,* 48, 21, 1983.
6. **Bussey, H. and Umbarger, H. E.,** Biosynthesis of branched-chain amino acids in yeast: regulation of synthesis of the enzymes of isoleucine and valine biosynthesis, *J. Bacteriol.,* 98, 623, 1969.
7. **Magee, P. T. and Hereford, L. M.,** Multivalent repression of isoleucine-valine biosynthesis in *Saccharomyces cerevisiae, J. Bacteriol.,* 98, 857, 1969.
8. **Freundlich, M., Burns, R. O., and Umbarger, H. E.,** Control of isoleucine, valine and leucine biosynthesis. I. Multi-valent repression, *Proc. Natl. Acad. Sci. U.S.A.,* 48, 1804, 1962.
9. **Bollon, A. P.,** Analysis of *ilv1 CIS* control and domain mutants, *Mol. Gen. Genet.,* 177, 283, 1980.
10. **Petersen, J. G. L., Holmberg, S., Nilsson-Tillgren, T., and Kielland-Brandt, M. C.,** Molecular cloning and characterization of the threonine deaminase *(ILV1)* gene of *Saccharomyces cerevisiae, Carlsberg Res. Commun.,* 48, 149, 1983.
11. **Polaina, J.,** Cloning of the *ILV2, ILV3* and *ILV5* genes of *Saccharomyces cerevisiae, Carlsberg Res. Commun.,* 49, 577, 1984.
12. **Kielland-Brandt, M. C., Holmberg, S., Petersen, J. G. L., and Nilsson-Tillgren, T.,** Nucleotide sequence of the gene for threonine deaminase *(ILV1)* of *Saccharomyces cerevisiae, Carlsberg Res. Commun.,* 49, 567, 1984.
13. **Petersen, J. G. L. and Holmberg, S.,** The *ILV5* gene of *Saccharomyces cerevisiae* is highly expressed, *Nucleic Acids Res.,* in press.
14. **Falco, S. C., Dumas, K. S., and Livak, K. J.,** Nucleotide sequence of the yeast *ILV2* gene which encodes acetolactate synthase, *Nucleic Acids Res.,* 13, 4011, 1985.
15. **Nilsson-Tillgren, T., Gjermansen, C., Kielland-Brandt, M. C., Petersen, J. G. L., and Holmberg, S.,** Genetic differences between *Saccharomyces carlsbergensis* and *S. cerevisiae.* Analysis of chromosome III by single chromosome transfer, *Carlsberg Res. Commun.,* 46, 65, 1981.
16. **Holmberg, S.,** Genetic differences between *Saccharomyces carlsbergensis* and *S. cerevisiae.* II. Restriction endonuclease analysis of genes in chromosome III, *Carlsberg Res. Commun.,* 47, 233, 1982.
17. **Nilsson-Tillgren, T., Gjermansen, C., Holmberg, S., Petersen, J. G. L., and Kielland-Brandt, M. C.,** Analysis of chromosome V and the *ILV1* gene from *Saccharomyces carlsbergensis, Carlsberg Res. Commun.,* 51, 309, 1986.
18. **Miozzari, G., Niederberger, P., and Hütter, R.,** Tryptophan biosynthesis in *Saccharomyces cerevisiae*: control of the flux through the pathway, *J. Bacteriol.,* 134, 48, 1978.
19. **Hinnebusch, A. G. and Fink, G. R.,** Repeated DNA sequences upstream from *HIS1* also occur at several other co-regulated genes in *Saccharomyces cerevisiae, J. Biol. Chem.,* 258, 5238, 1983.
20. **Hinnebusch, A. G. and Fink, G. R.,** Positive regulation in the general amino acid control of *Saccharomyces cerevisiae, Proc. Natl. Acad. Sci. U.S.A.,* 80, 5374, 1983.
21. **Fink, G. R.,** Gene-enzyme relations in histidine biosynthesis in yeast, *Science,* 146, 525, 1964.
22. **Wolfner, M., Yep, D., Messenguy, F., and Fink, G. R.,** Integration of amino acid biosynthesis into the cell cycle of *Saccharomyces cerevisiae, J. Mol. Biol.,* 96, 273, 1975.
23. **Niederberger, P., Miozzari, G., and Hütter, R.,** Biological role of the general control of amino acid biosynthesis in *Saccharomyces cerevisiae, Mol. Cell. Biol.,* 1, 584, 1981.
24. **Falco, S. C. and Dumas, K. S.,** Genetic analysis of mutants of *Saccharomyces cerevisiae* resistant to the herbicide sulfometuron methyl, *Genetics,* 109, 21, 1985.
25. **Holmberg, S., Kielland-Brandt, M. C., Nilsson-Tillgren, T., and Petersen, J. G. L.,** The *ILV1* gene in *Saccharomyces cerevisiae*: 5′ and 3′ end mapping of transcripts and their regulation, *Carlsberg Res. Commun.,* 50, 163, 1985.
26. **Neupert, W. and Schatz, G.,** How proteins are transported into mitochondria, *TIBS,* 76, 1, 1981.
27. **Conde, J. and Fink, G. R.,** A mutant of *Saccharomyces cerevisiae* defective for nuclear fusion, *Proc. Natl. Acad. Sci. U.S.A.,* 73, 3651, 1976.
28. **Nilsson-Tillgren, T., Petersen, J. G. L., Holmberg, S., and Kielland-Brandt, M. C.,** Transfer of chromosome III during *kar* mediated cytoduction in yeast, *Carlsberg Res. Commun.,* 45, 113, 1980.

29. **Dutcher, S. K.,** Internuclear transfer of genetic information in *kar1-1/KAR1* heterokaryons in *Saccharomyces cerevisiae, Mol. Cell. Biol.,* 1, 245, 1981.
30. **Holmberg, S. and Petersen, J. G. L.,** Regulation of isoleucine biosynthesis in *Saccharomyces cerevisiae,* submitted.
31. **de Robichon-Szulmajster, H. and Magee, P. T.,** The regulation of isoleucine-valine biosynthesis in *Saccharomyces cerevisiae.* I. Threonine deaminase, *Eur. J. Biochem.,* 3, 492, 1968.
32. **Miozzari, G. F., Niederberger, P., and Hütter, R.,** Permeabilization of microorganisms by Triton X-100, *Anal. Biochem.,* 90, 220, 1978.
33. **Fink, G. R. and Styles, C. A.,** Gene conversion of deletions in the *HIS4* region of yeast, *Genetics,* 77, 231, 1966.
34. **Gjermansen, C.,** unpublished.

Chapter 16

THE DEVELOPMENT OF CELLULOLYTIC YEASTS AND THEIR POSSIBLE APPLICATIONS

Jonathan Knowles, Merja Penttilä, Tuula Teeri, Lars André, Päivi Lehtovaara, and Irma Salovuori

TABLE OF CONTENTS

I. INTRODUCTION

Yeasts are used widely in industry, particularly for wine and beer production, bread baking, ethanol production, the production of single cell protein, and the production of vitamins and fine chemicals. Many of the raw materials or feedstocks used in these processes are of plant origin and often contain significant amounts of cellulose.

In general, yeasts are efficient utilizers of a range of different soluble saccharides and some can degrade starch. However, yeasts capable of degrading the insoluble components of lignocellulose-cellulose, hemicellulose, and lignin have not been described. Certain filamentous fungi, such as the well-studied *Trichoderma reesei,* secrete large quantities of a mixture of different cellulolytic enzymes into the growth medium, and this mixture efficiently hydrolyzes crystalline cellulose to glucose.[1-3] Certain bacteria such as *Clostridium* and *Cellulomonas* also secrete cellulases into the medium, although these enzymes are often bound as a complex to the cell wall.[4]

Recombinant DNA technology makes possible the transfer of genes between different organisms in a specific manner and so permits the construction of novel organisms with completely new properties. The purpose of this paper is to review recent attempts, both by ourselves and others, to construct yeast strains producing cellulolytic enzymes and then to discuss the possible use of this type of recombinant organism in different industrial processes.

II. CELLULOLYTIC ENZYMES

A. Properties

Cellulose, the major constituent of plant cell walls, is composed of long chains of β-1,4-glucan. Cellulose can form very compact crystalline regions which are relatively refractory to enzymatic hydrolysis. The alternation between amorphous and crystalline regions leads to a variable structure of this chemically simple substrate.[5] The brown rot fungus *Trichoderma reesei* is probably the best-studied microbial producer of cellulases. An efficient hydrolysis of cellulose requires a number of different enzymes, but it is not yet possible to define the roles of individual enzymes in this hydrolysis. Nevertheless, despite certain differences of opinion as to the substrate specificity of particular fungal enzymes, it is possible to divide the activities into two classes: endo- and exoglucanases.[1-3] Endoglucanases (EC 3.2.1.4) will cleave randomly within a β-1,4-glucan molecule, and modified soluble substrates such as hydroxyethyl or carboxymethyl cellulose are efficiently cleaved.[3,6,8-12] Exoglucanases act on the ends of the molecule. A typical example of this type would be cellobiohydrolases (EC 3.2.1.91), which cleave cellobiose from the nonreducing end of the β-glucan molecule. Interestingly, it has been found that two different cellulases, acting together on a given substrate, can have an activity which is more than twice the sum of activities of the individually acting enzymes.[6-8,10] This phenomenon, synergy, has been observed between both endo-exo enzymes and between different exo-enzymes.

Each enzyme probably plays a specific role in the complete hydrolysis and the importance of any particular role will depend very much on the substrate to be hydrolyzed. For applications involving the use of grain as raw material such as brewing and ethanol production, an enzyme degrading β-glucans such as *T. reesei* EG I is required. For applications involving cellulosic raw materials such as lignocellulose, enzymes degrading crystalline cellulose such as *T. reesei* CBH I and CBH II are also necessary. It is also evident that other properties of the enzyme are important and will influence the choice. For instance, many fungal cellulases have relatively low pH optima between 3 and 5, whereas β-glucanases from Bacilli often have optimi around pH 7.[2] The particular gene to be transferred into an industrial host strain obviously should code for an enzyme which can function under the conditions required.

B. Cloning of Cellulase Genes

With the long-term goal of increasing our understanding of the enzymatic hydrolysis of cellulose, we have isolated and characterized a number of genes coding for cellulolytic enzymes from filamentous fungi. The first gene we were able to isolate was a β-glucosidase gene from *Aspergillus niger*.[13] It was earlier thought that yeast and filamentous fungi were so closely related that fungal genes might be expressed in yeast. With this idea in mind we constructed a gene bank from *A. niger* in a yeast cosmid vector and screened for the expression of five different fungal genes in the yeast *Saccharomyces cerevisiae*. Only one of these, β-glucosidase, was expressed and this at a low level. Nevertheless, yeast strains containing this fungal gene were able to grow slowly on the disaccharide cellobiose, one of the major hydrolysis products of cellulose.

The experiments just described were of use in that they showed that cloning of fungal genes by expression in yeast was clearly not an efficient method for the isolation of specific fungal genes. For this reason, differential hybridization of radioactive cDNAs was used to identify genes from *Trichoderma reesei* strongly expressed during cellulase synthesis.[14]

All four of the genes coding for the major cellulases of *T. reesei* were isolated from λ-phage gene banks by differential hybridization using two populations of cDNA, made from either induced (cellulase-positive) or repressed (cellulase-negative) mRNA. Hybrid phages containing genes coding for the cellobiohydrolases CBH I and CBH II, and the endoglucanases EG I and EG II were identified using hybrid selection of message and immunoprecipitation of in vitro synthesized products. Also, full length cDNA copies corresponding to these genes were isolated, and both the chromosomal genes and cDNAs for CBH I, CBH II, and EG I have been sequenced.[14-17] CBH I and EG I show over 40% amino acid homology to each other, but almost none to CBH II. However, CBH II and EG II also seem to be related to each other.[16,18] A group working at the Cetus Corporation has also isolated and characterized genes from *T. reesei* coding for CBH I and EG I,[19,20] and their results concerning these two genes agree substantially with those that we have obtained. An endoglucanase gene from another fungus, *Schizophyllum commune,* has also been isolated.[21]

A number of bacterial cellulase genes have been isolated by several different groups. The isolation of bacterial genes appears to be much easier than the isolation of fungal cellulase genes. In most cases it is possible to isolate the gene by direct screening of gene banks made in *Escherichia coli* for an endoglucanase type of activity on a suitable medium such as carboxymethyl cellulose. Endoglucanase genes from *Thermomonospora,*[22] *Clostridium,*[23] *Cellulomonas,*[24] and *Bacillus* species[25-27] have so far been isolated.

Thus, a number of different cloned cellulase genes both from fungi and from bacteria are now available for the construction of cellulolytic yeasts.

III. CONSTRUCTION OF CELLULOLYTIC YEASTS

A. Expression of a Full Length CBH I cDNA in Yeast

Clearly, to obtain efficient expression of fungal cellulase genes in yeast it is necessary to replace the fungal promotor and terminator sequences with a strong yeast promoter and yeast terminator. Since recent reports suggest that *Saccharomyces* is unable to splice efficiently fungal introns,[28] and since the internal yeast splicing signal, TACTAAC,[29] is not found within the introns of any of the fungal genes studied here, full length fungal cDNAs were used as the coding sequences. As shown in Figure 1, plasmid pMP11 was constructed by inserting a full length cDNA copy of CBH I into the phosphoglycerokinase expression cassette of plasmid pMA 91 kindly provided by Dr. Alan Kingsman.[30]

Yeast cells containing plasmid pMP11 were grown up on minimal medium, and the growth medium without cells was concentrated by ultrafiltration. The secreted proteins were analyzed on SDS-PAGE followed by western blotting (Figure 2). These data show that CBH I produced

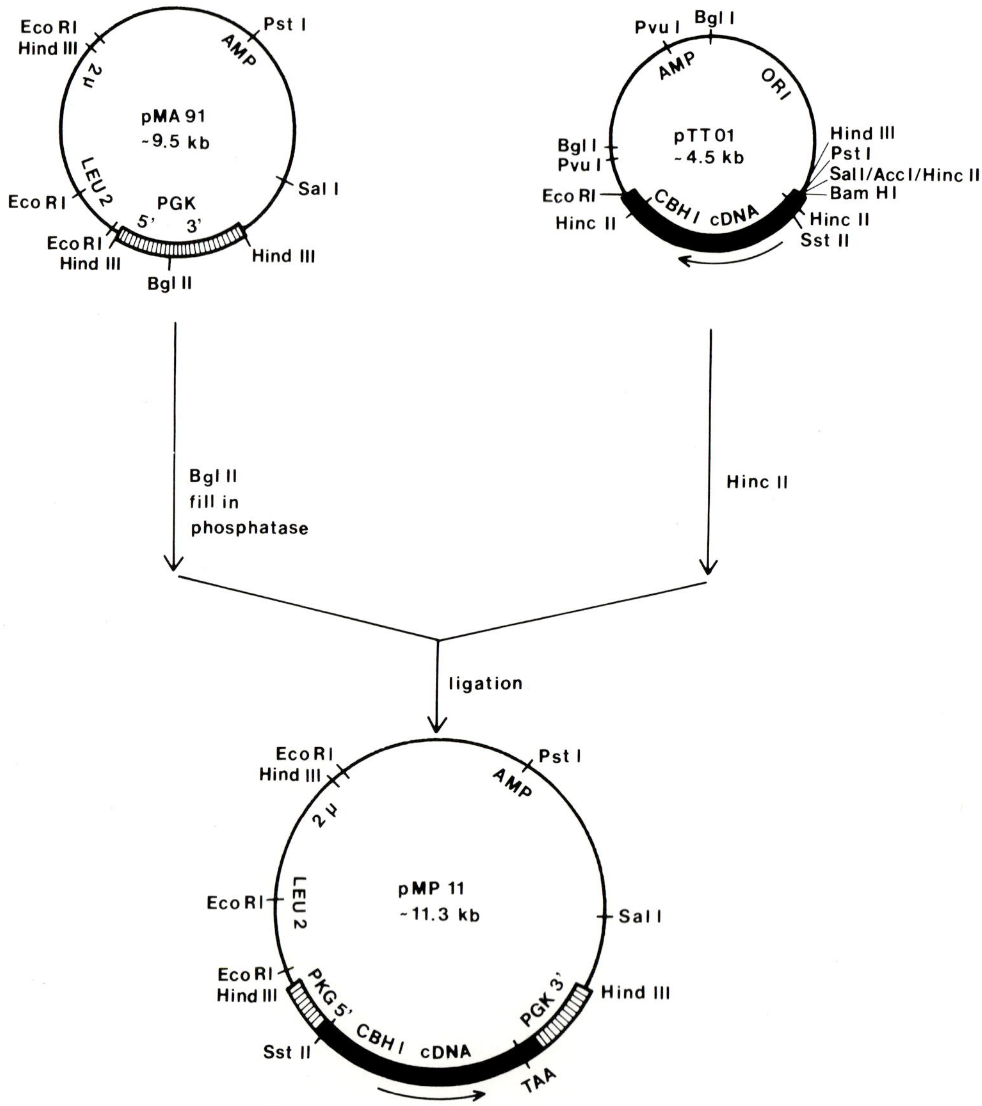

FIGURE 1. This shows diagrammatically the construction of plasmid pMP 11 for the expression of the CBH I cDNA in *Saccharomyces cerevisiae*.

in yeast from this cDNA construction (lane a) is secreted into the medium in a more heavily glycosylated form than the authentic enzyme which has a molecular weight of 68,000 (lane d). This overglycosylation can be removed by digestion with Endo H (lanes b and c), indicating that the carbohydrate is attached to the protein by N-linkages. The amount of CBH I found in the medium at stationary phase in minimal media corresponds to about 1 mg/ℓ or approximately 0.5% of total cell protein. These results suggest that the CBH I fungal signal sequence is recognized by the yeast secretory system as has been shown for amyloglucosidase.[28]

The Cetus Corporation has also expressed a CBH I intronless gene from *Trichoderma reesei* in yeast and has obtained similar results to those described above.[20] With some optimization of the mRNA leader sequence, the Cetus group was able to obtain around 10 mg/ℓ of secreted protein at stationary phase. It is, however, likely that the precise growth conditions will significantly affect the amount of cellulase secreted into the medium.

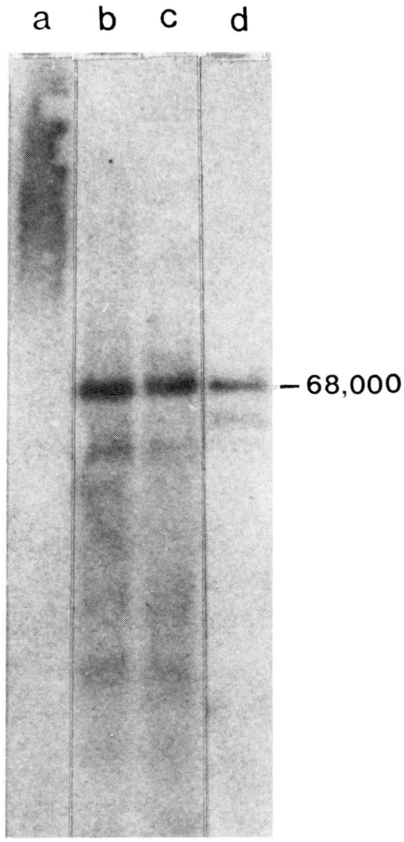

FIGURE 2. Proteins from the growth medium of yeast containing pMP 11 and control yeast were concentrated by ultrafiltration and analyzed by SDS PAGE and western blotting with antiserum against native CBH I. Lane (a) shows the high molecular weight product found in the growth medium of the yeast containing pMP 11 that reacted with anti-CBH I antibodies. Lane (d) shows the native protein of molecular weight 68,000 produced by *T. reesei* and lanes (b) and (c) show the yeast-synthesized CBH I treated with Endo H for 24 and 72 hr, respectively, without prior denaturation.

B. Expression of a Full Length EG I cDNA in Yeast

The construction of plasmid used for the expression of EG I cDNA, the major endoglucanase of *T. reesei,* was done essentially as described for CBH I (Figure 3). EG I was also secreted by yeast with the aid of its own signal peptide. The secreted EG I was in an overglycosylated form, but, nevertheless, proved to be active on mixed 1,3-1,4 β-glucans. Figure 4 shows the hydrolysis of lichenin, a 1,3-1,4 β-glucan, by a yeast strain carrying the EG I cDNA in the PGK expression cassette. The hydrolysis of lichenin on minimal agar plates is revealed by staining with Congo red. The control strain of yeast did not hydrolyze the lichenin. Similar results with the expression of an intronless EG I gene have been obtained by the Cetus Corporation.[20]

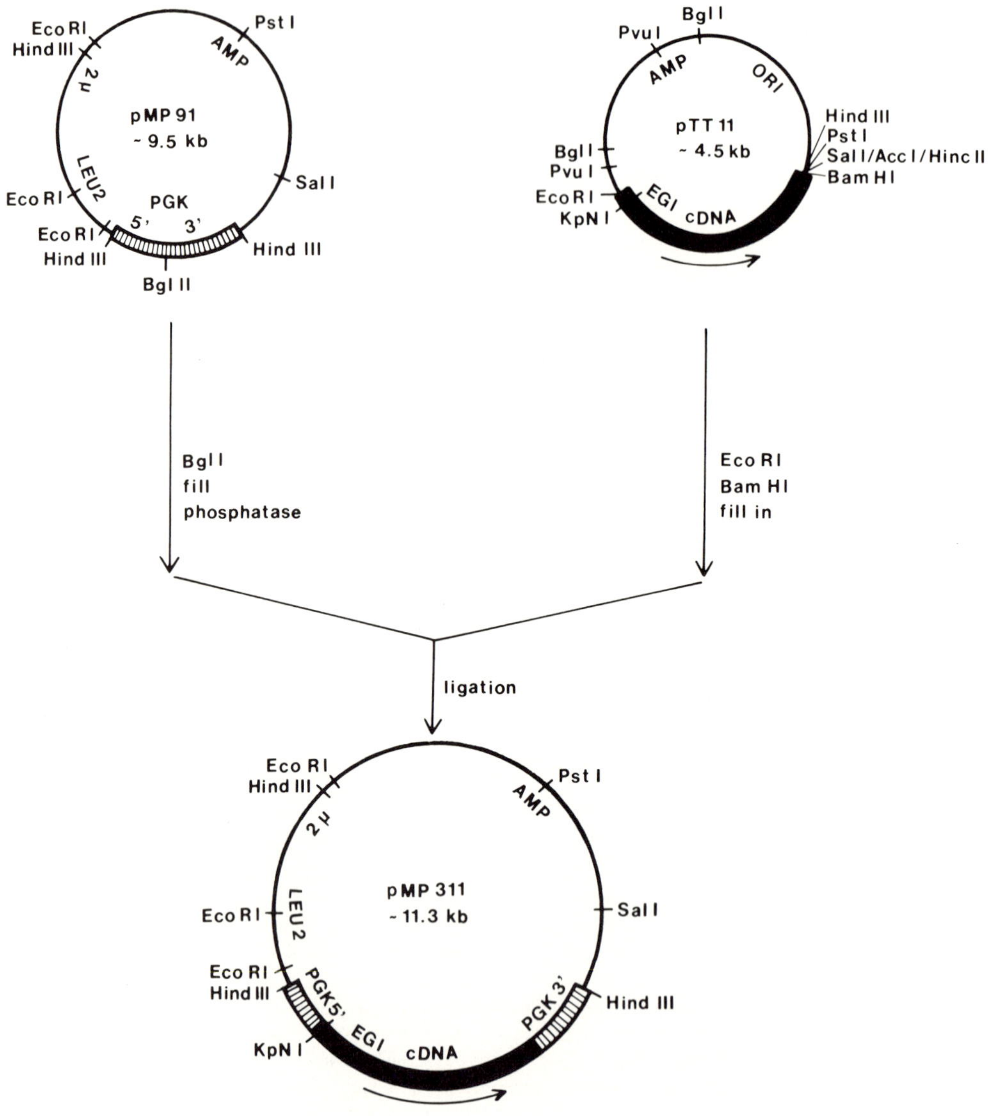

FIGURE 3. This shows diagrammatically the construction of plasmid pMP 311 for the expression of the EG I cDNA in *Saccharomyces cerevisiae*.

C. Expression of a Full Length CBH II cDNA in Yeast

The construction used for the expression of CBH II in yeast was made in a similar way to that used for EG I expression as shown in Figure 3. Yeasts containing CBH II cDNA under the control of the yeast PGK promoter also secrete CBH II into the growth medium. Figure 5 shows that CBH II produced by yeast is able to hydrolyze lichenin. In view of the current uncertainty concerning the precise activity of purified cellulases, this finding, not described before, is of interest. It would seem that CBH II, like CBH I,[7] may also have some endoglucanase activity on β 1,3-1,4 glucan. Further work is being done to characterize the efficiency of secretion and the biochemistry of these fungal cellulases produced in yeast.

In summary, work at VTT and also at the Cetus Corporation has shown that yeast will produce and secrete a number of different fungal cellulases capable of degrading β-glucans.

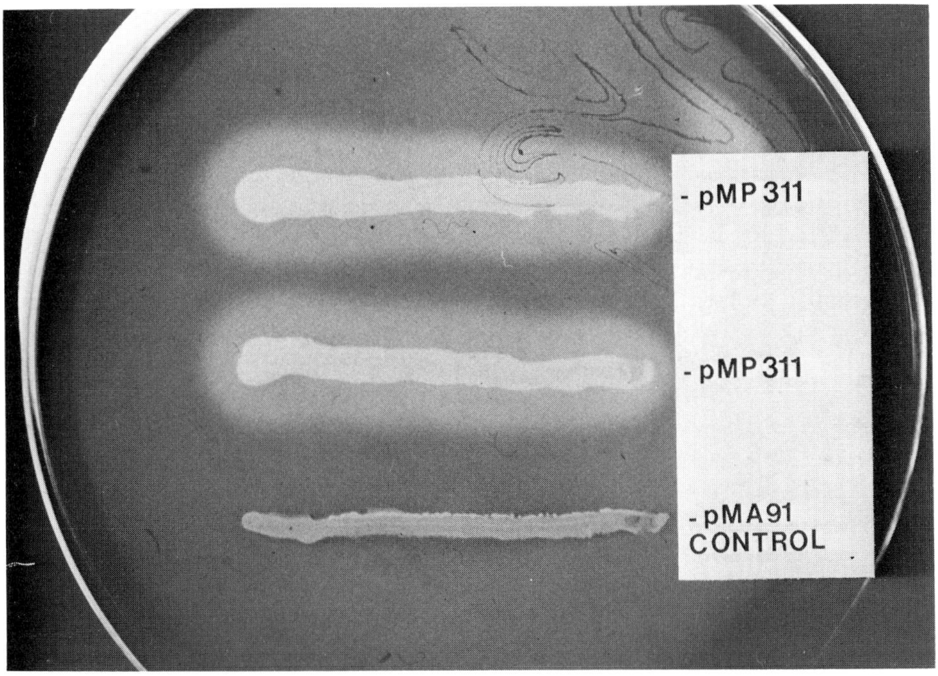

FIGURE 4. Two strains of yeast, one containing plasmid pMP 311 for the expression of EG I and a control, were grown for 2 days on minimal medium agar containing 0.5% Lichenin. The extent of hydrolysis of lichenin was then revealed by staining with Congo red.

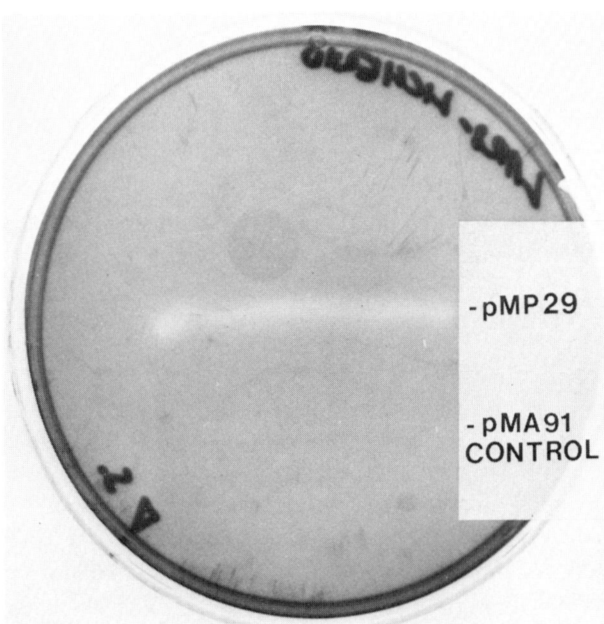

FIGURE 5. Two strains of yeast, one containing plasmid pMP 29 for the expression of CBH II and a control, were grown on lichenin minimal medium plates as for Figure 4. The extent of hydrolysis was revealed by staining with Congo red.

The amounts produced are, of course, relatively small compared to those produced by the original donor organism, *T. reesei,* being of the order of 1% or less of total cell protein, but for a number of commercial applications this may prove sufficient.

D. Expression of Bacterial Cellulases in Yeast

All reports of bacterial cellulase genes cloned so far concern genes coding for enzymes showing endo-type activity, that is, hydrolysis of mixed glucans such as lichenin and barley β-glucan, and modified celluloses such as carboxymethyl cellulase. Two groups have transferred the endo-1,3-1,4-β-glucanase gene from *Bacillus subtilis* to yeast. Hinchliffe and Box[31] transferred the intact *Bacillus* gene to yeast and observed some β-glucanase activity, all located intracellularly. In contrast, Cantwell et al.[32] were able to detect enzyme activity in the culture medium by using a yeast ADH promotor to express the bacterial glucanase coding sequences. Whether this difference is due to the increased level of expression or a modification to the *Bacillus* coding sequence is not yet clear. Cantwell et al.[32,33] have also transferred the β-glucanase expression cassette to different brewing strains and found that these new recombinant strains were also able to produce significant amounts of the bacterial enzyme.

Sacco et al.[34] have reported that an endoglucanase gene from *Clostridium thermocellum* is expressed in yeast, but that the enzyme activity is found in the cytoplasm despite the fact that this enzyme is secreted in both *Clostridium* and *Escherichia coli.* Skipper et al.[35] have been able to express and secrete a cellulase from *Cellulomonas fimi* by fusing the gene to the yeast MEL I gene.

In summary, it would seem that some bacterial cellulases can also be secreted by yeast if the bacterial structural genes are joined to efficient yeast regulatory and secretory sequences. Transfer of unmodified bacterial cellulase genes to yeast may give enzyme activity, but this is often localized inside the cell.

IV. POSSIBLE APPLICATIONS OF CELLULOLYTIC YEASTS

The extreme specificity of recombinant DNA technology should allow us to add to an industrial strain just those properties we require, such as the ability of utilizing a novel substrate, without in any way modifying other desirable properties which may have been accumulated over long periods of selection and screening. Insertion of cellulase genes into an industrial strain should not, at least in principle, reduce its tolerance to ethanol. Before constructing cellulolytic yeasts for a specific application, it is important to define the substrate in terms of the particular enzymes required.

A. The Brewing Industry

Roughly 10% of the total carbohydrate of barley grain is β-glucan. Much of this is hydrolyzed during the malting and mashing processes by barley enzymes. Excess β-glucan in beer is undesirable, because it can have a dramatic effect on beer filtration and may cause problems with beer clarity. The amount of β-glucan remaining depends on factors such as the nature of the growing season and the precise conditions of malting.[36] In addition, the use of adjuncts, unmalted grain, or other material can significantly raise the β-glucan content of the wort to be fermented. For these reasons, some breweries routinely add β-glucanases to the wort before fermentation.[37]

Brewing yeast secreting fungal or bacterial β-glucanase could well find practical application in the not-too-distant future. The construction of a stable brewing strain secreting β-glucanase by integration of a glucanase gene into the chromosome clearly should not pose any insurmountable problems. It is, however, desirable that no vector sequences are transferred into a strain to be used for production. For brewing applications, the required level of expression of an endoglucanase is low, much less than 1% of total cell protein, provided that the enzyme has a low pH optimum.[36,37]

This type of endoglucanase-producing brewing yeast should provide the following benefits: first, beer filtration should be easier and cheaper with less frequent cleaning of filters. This could also improve the production of filter-sterilized draft beers which cannot be pasteurized. Second, the brewing process would better tolerate barley with high glucan content and the addition of unmalted adjuncts. Similar types of problems also occur in the wine-making industry and the construction and use of a β-glucanase-secreting wine yeast might well benefit certain types of wine production.

B. Ethanol Production

The sucrose or glucose used for the production of ethanol by yeast fermentation is most commonly obtained from plant starch material, which is hydrolyzed enzymatically before the fermentation process. The fermentation feedstock will contain a variable amount of β-glucan, some of it as cellulose or lignocellulose and some of it as mixed linkage β-glucans. Utilization of this carbohydrate would increase the efficiency of the fermentation, if other parameters remain unaffected.[38]

Continuous production and secretion of cellulase during fermentation of yeast at high cell density could lead to useful amounts of enzyme in the medium. Many cellulases are stable under the conditions of fermentation for a number of days. If it proved possible to develop a really efficient celulase-secreting yeast, the fermentation feedstock could be augmented with lignocellulose to make the process even more economical.

In some countries, e.g., Brazil,[39] manufacture of ethanol by fermentation is a major industry. Where the volume of a product is high, even very small improvements in efficiency lead to substantial savings. It is in this context that cellulolytic yeast could make a contribution to improving the efficiency of ethanol fermentation.

C. Utilization of Lignocellulose

The enzymatic hydrolysis of lignocellulose to give sugar which could then be used to produce ethanol, single cell protein, or other useful products has been the subject of much discussion in recent years.[40,41] There are a number of problems to be overcome before this type of process becomes viable. One of the more important is the cost of the cellulolytic enzymes. The information available at present suggests that it may prove very difficult, if not impossible, to engineer yeast to secrete cellulases as efficiently as a fungus like *T. reesei*. For this reason, a one-step hydrolysis-fermentation process of lignocellulose by cellulolytic yeast seems very unlikely ever to be profitable.

In conclusion, it is now possible to transfer genes coding for different cellulases into yeast, and so produce organisms which can degrade cellulose and other β-glucans and ferment the sugars so produced. The first applications of this type of organism will most likely be in the brewing industry and, perhaps, later in the wine-making industry. When more information concerning the nature of the cellulose and other β-glucans present in feedstocks is available, it should be possible to improve the efficiency of large-scale ethanol fermentations by constructing specific recombinant yeasts for particular processes. In general, much more information is needed concerning the mechanism of action of individual cellulases in cellulase hydrolysis before these important enzymes can be exploited to the fullest.

ACKNOWLEDGMENTS

This research was funded in part by Alko Ltd., Finnish State Alcohol Company, and the National Swedish Board for Technology.

REFERENCES

1. **Gritzali, M. and Brown, R. D., Jr.,** The cellulose system of *Trichoderma, Adv. Chem. Ser.*, 181, 237, 1979.
2. **Enari, T.-M.,** Microbial cellulases, in *Microbial Enzymes and Biotechnology*, Fogarty, W. M., Ed., Applied Science Publishers, London, 1983, 183.
3. **Montenecourt, B. S.,** *Trichoderma reesei* cellulases, *Trends Biotechnol.*, 2, 156, 1983.
4. **Bayer, E. A., Setter, E., and Lamed, R.,** Organization and distribution of the cellulosome in *Clostridium thermocellum, J. Bacteriol.*, 163, 552, 1985.
5. **Sjöström, E.,** *Wood Chemistry, Fundamentals and Applications*, Academic Press, New York, 1981.
6. **Henrissat, B. et al.,** Synergism of cellulases from *Trichoderma reesei* in the degradation of cellulose, *Bio/technology*, 3, 722, 1985.
7. **Nummi, M. et al.,** Cellobiohydrolase from *Trichoderma reesei, Biochem. J.*, 215, 677, 1983.
8. **Wood, T. M.,** Enzyme interactions involved in fungal degradation of cellulosic materials, in The Ekman-Days, Int. Symp. on Wood and Pulping Chemistry, Stockholm, 1981.
9. **Beldman, G. et al.,** The cellulase of *Trichoderma viride, Eur. J. Biochem.*, 146, 301, 1985.
10. **Pettersson, G. et al.,** The cellulase complex of *Trichoderma reesei* QM 9414, in The Ekman Days, Int. Symp. on Wood and Pulping Chemistry, Stockholm, 1981.
11. **Shoemaker, S. et al.,** Characterization and properties of cellulases purified from *Trichoderma reesei* strain L27, *Bio/technology*, 1, 689, 1983.
12. **Niku-Paavola, M.-L. et al.,** A new appraisal of the endoglucanases of the fungus *Trichoderma reesei, Biochem. J.*, in press.
13. **Penttilä, M. E. et al.,** Cloning of *Aspergillus niger* genes in yeast, *Mol. Gen. Genet.*, 194, 494, 1984.
14. **Teeri, T., Salovuori, I., and Knowles, J.,** The molecular cloning of the major cellulase gene from *Trichoderma reesei, Bio/technology*, 1, 696, 1983.
15. **Penttilä, M. et al.,** The isolation and sequence of the gene coding for the major endoglucanase of *Trichoderma reesei,* submitted.
16. **Teeri, T. et al.,** The cloning and isolation of a full length cDNA coding for cellobiohydrolase II from *Trichoderma reesei,* submitted.
17. **Lehtovaara, P. et al.,** The structure of the chromosomal gene coding for the cellobiohydrolase II from *Trichoderma reesei,* in preparation.
18. **Saloheimo, M.,** unpublished.
19. **Shoemaker, S. et al.,** Molecular cloning of exo-cellobiohydrolase I derived from *Trichoderma reesei* strain L27, *Bio/technology*, 1, 691, 1983.
20. **Shoemaker, S. et al.,** Recombinant Fungal Cellobiohydrolases, European Patent Appl. 84 110, 305.4, 1984.
21. **Seligy, V.,** unpublished.
22. **Collmer, A. and Wilson, D. B.,** Cloning and expression of a *Thermomonospora Yx* endoglucanase gene in *E. coli, Bio/technology*, 1, 594, 1983.
23. **Cornet, P. et al.,** Characterization of two *cel* (cellulose degradation) genes from *Clostridium thermocellum* coding for endoglucanases, *Bio/technology*, 1, 589, 1983.
24. **Gilkes, N. et al.,** Isolation and characterization of *Escherichia coli* clones expressing cellulase genes from *Cellulomonas fimi, J. Gen. Microbiol.*, 130, 1377, 1984.
25. **Cantwell, B. and McConnell, D.,** Molecular cloning and expression of a *Bacillus subtilis* β-glucanase gene in *Escherichia coli, Gene*, 23, 211, 1983.
26. **Hinchliffe, E.,** Cloning and expression of a *Bacillus subtilis* endo-1,3-1,4-β-D-glucanase gene in *Escherichia coli* K12, *J. Gen. Microbiol.*, 130, 1285, 1984.
27. **Rainer, B., Bäumlein, H., and Hofemeister, J.,** Expression in *Escherichia coli* of a cloned β-glucanase gene from *Bacillus amyloliquefaciens, Appl. Microbiol. Biotechnol.*, 22, 63, 1985.
28. **Innis, M. et al.,** Expression, glycosylation, and secretion of an *Aspergillus* glucoamylase by *Saccharomyces cerevisiae, Science*, 228, 21, 1985.
29. **Langford, C. and Gallwitz, D.,** Evidence for an intron-contained sequence required for the splicing of yeast RNA polymerase II transcripts, *Cell*, 33, 519, 1983.
30. **Mellor, J. et al.,** Efficient synthesis of enzymatically active calf chymosin in *Saccharomyces cerevisiae, Gene*, 24, 1, 1983.
31. **Hinchliffe, E. and Box, W.,** Expression of the cloned endo-1,3-1,4-β-glucanase gene of *Bacillus subtilis* in *Saccharomyces cerevisiae, Curr. Genet.*, 8, 471, 1984.
32. **Cantwell, B. et al.,** Expression of the gene for the endo-β-1,3-1,4-glucanase from *Bacillus subtilis* in *Saccharomyces cerevisiae,* in Proc. 20th Congr. of the European Brewery Convention, 1985, 259.
33. **Cantwell, B. A. and McConnell, D. J.,** Transformation of brewing yeast strains, in *Biological Research on Industrial Yeasts*, Vol. 2, Stewart, G. G., Russell, I., Klein, R. D., and Kiebsch, R. R., Eds., CRC Press, Boca Raton, Fla., 1987, chap. 8.

34. **Sacco, M., Millet, J., and Aubert, J. P.,** Cloning and expression in *Saccharomyces cerevisiae* of a cellulase gene from *Clostridium thermocellum, Ann. Microbiol. (Inst. Pasteur),* 135A, 485, 1984.
35. **Skipper, N. et al.,** Use of the melibiase promoter and signal peptide to express a bacterial cellulase from yeast, in *Biological Research on Industrial Yeasts,* Vol. 1, Stewart, G. G., Russell, I., Klein, R. D., and Hiebsch, R. R., Ed., CRC Press, Boca Raton, Fla., 1987, chap. 11.
36. **Knowles, J. et al.,** Transfer of genes coding for fungal glucanses into yeast, in Proc. 20th Congr. of the European Brewery Convention, 1985, 251.
37. **Hinchliffe, E.,** β-Glucanase: the successful application of genetic engineering, *J. Inst. Brewing,* 91, 384, 1985.
38. **Oksanen, J.,** Using cellulase in grain alcohol fermentation, in Proc. Biotech '84 Europe, 1984, 637.
39. **Tavares, F. C. A. and Echeverrigaray, S.,** Yeast breeding for fuel production, in *Biological Research on Industrial Yeasts,* Vol. 1, Stewart, G. G., Russell, I., Klein, R. D., and Hiebsch, R. R., Eds., CRC Press, Boca Raton, Fla., 1987, chap. 5.
40. **Ng, T. et al.,** Production of feedstock chemicals, *Science,* 219, 733, 1983.
41. **Litchfield, J. H.,** Single cell proteins, *Science,* 219, 740, 1983.

INDEX

D

E

F

N

O

P

T

U

V

W

X

Y

Z